Impedance
Microbiology

INNOVATION IN MICROBIOLOGY SERIES

Series Editor: **Dr A. N. Sharpe,** *Health and Welfare, Ottawa, Canada*

1. The Direct Epifluorescent Filter Technique for the rapid enumeration of micro-organisms
   G. L. Pettipher

2. Utilization of Microorganisms in Meat Processing: A handbook for meat plant operators
   Jim Bacus

3. Impedance Microbiology
   Ruth Eden *and* **Gideon Eden**

Impedance Microbiology

R. Firstenberg-Eden, Ph.D.
and
G. Eden, Ph.D.
BACTOMATIC,
*A Division of Medical Technology Corporation,
Princeton, New Jersey, United States of America*

RESEARCH STUDIES PRESS LTD.
Letchworth, Hertfordshire, England
JOHN WILEY & SONS INC.
New York · Brisbane · Chichester · Toronto · Singapore

RESEARCH STUDIES PRESS LTD.
58B Station Road, Letchworth, Herts. SG6 3BE, England

Copyright © 1984, by Research Studies Press Ltd.

All rights reserved.

No part of this book may be reproduced by any means, nor transmitted, nor translated into a machine language without the written permission of the publisher.

Marketing and Distribution:

Australia, New Zealand, South-east Asia:
Jacaranda-Wiley Ltd., Jacaranda Press
JOHN WILEY & SONS INC.
GPO Box 859, Brisbane, Queensland 4001, Australia

Canada:
JOHN WILEY & SONS CANADA LIMITED
22 Worcester Road, Rexdale, Ontario, Canada

Europe, Africa:
JOHN WILEY & SONS LIMITED
Baffins Lane, Chichester, West Sussex, England

North and South America and the rest of the world:
JOHN WILEY & SONS INC.
605 Third Avenue, New York, NY 10158, USA

Library of Congress Cataloging in Publication Data:

Firstenberg-Eden, R. (Ruth), 1943–
  Impedance microbiology.
  (Innovation in microbiology series; 3)
  Bibliography: p.
  Includes index.
  1. Micro-organisms—Electric properties—Measurement.
2. Impedance, Bioelectric—Measurement. 3. Microbial growth—Measurement. I. Eden, G. (Gideon), 1945–
II. Title. III. Series.
QR69.E43F57   1984   576'.119127   84-17947
ISBN 0 471 90623 9 (Wiley)

British Library Cataloguing in Publication Data:

Firstenberg-Eden, R.
  Impedance microbiology.—(Innovations in microbiology series; 3)
  1. Impedance, Bioelectric   2. Microbial metabolism
  I. Title   II. Eden, G.   III. Series
  576'.119127   QP341

ISBN 0 86380 020 3

ISBN 0 86380 020 3 (Research Studies Press Ltd.)
ISBN 0 471 90623 9 (John Wiley & Sons Inc.)

Printed in Great Britain

Preface

This book was written to provide microbiologists and researchers, involved in rapid microbial measurement methods, with a comprehensive and up-to-date account of basic principles and advanced techniques of impedance microbiology.

Initial efforts to study electrical impedance variations resulting from the metabolic activity of growing microorganisms may be traced to the late 1890's when G. N. Stewart described, before the British Medical Association, measurements of conductivity depression on samples of defibrinated dog blood allowed to putrefy. It took nearly a century to improve upon and implement the basic ideas in a commercial type instrument for rapid estimation of bacterial concentrations.

Impedance microbiology has become increasingly recognized in the field of rapid bacteriological methods. Interest is still growing rapidly in the fields of clinical and food microbiology. Each field emphasizes certain aspects such as correlation with standard plate counts, shelf-life, sensitivity to antibiotic susceptibilities, etc.

The bulk of the material dealing with impedance microbiology is widely scattered amongst technical journals and conference proceedings. Consequently, it is a rather difficult task, particularly for a newcomer, to learn the principles underlying the subject. This text attempts to put between the covers of one book information on the fundamental principles and up-to-date techniques of impedance microbiology, and to organize it in a coherent and unified manner.

It is assumed that the reader has an adequate background in general microbiology. The "electro-bacteriological" model presented in chapter 3 and the statistical analysis presented in Appendix A require some background in mathematical analysis and statistics. Definitions of terms used throughout the text are provided in Appendix B. In presenting the material, emphasis is placed on the development of practical means to obtain reliable results in a variety of applications. Most of the theoretical and practical materials presented in this book are the results of research carried out over the last two years.

Acknowledgement

We would like to thank Joseph DeSanto for maintaining a research environment that made this book possible, for discussing the various theories and for his encouragement throughout the various stages of the book preparation. Special thanks to Jo Ann Kropilak who typed and organized this manuscript many times with a cheerful spirit, and to Keppler Castaño for his artistic work on the illustrations. We are indebted to a large number of individuals who helped us to collect the data presented and contributed to the better understanding of impedance microbiology, they include: Dr. Joseph Zindulis, Corbett Klein, Marylou Van Sise, Peter Kahn, Martin Tricarico and Annikka Rantama. Finally we are grateful to Dr. A.N. Sharpe whose experience has been a major contributor to the formation of this book.

Foreword

Few persons or events affected the course of a scientific discipline so profoundly as did Robert Koch and his publication in 1881 of an agar-based method for isolating bacteria. The track thus set for microbiology was cemented for the next century by Koch's assistant, Petri, with his description of '...a small modification to the master's apparatus.'.

Notwithstanding the development of numerous alternatives, most of the advances in analytical microbiology stem from a focussing on concepts of isolation, pure cultures, and colony forming units that the plating methods engendered. And microbiologists, in the main, have accepted prolonged incubation as an inescapable consequence of the plating procedure's sensitivity.

Today we cannot afford to wait so long to get results. Nor can we afford to centre attention so closely, merely on the properties of individual microorganisms or how many of them exist in the sample. Mixed populations are the rule rather than the exception and the interactions between microorganisms in a sample ensure that the whole does not behave like the sum of the parts. Usually it is how actively microbial contamination will do what it will do that matters, not how many of the individuals of a species are there.

The plating methods set a level of sensitivity that quite outperformed that of the early alternative methods. But limits of detection and analysis times of plate methods remained more or less constant, while physical and chemical techniques advanced rapidly. Thus it is, that most microbiologists must now consider abandoning the old and trusted tool in favor of more efficient modern instrumentation. And of the available techniques, electrical impedance must − through its rapidity, versatility, compatibility with automation and computerized control, and pertinence to the total metabolic activity of the organisms − be considered a most desirable approach.

One can sense the quickening interest in impedance in almost every branch of microbiology. But the subject has long lacked the reference base needed to acquaint newcomers to the field with the theories and potentials that will allow them to adapt the technique to their individual problems. This obstacle has now been removed. With this book, Drs. Ruth and Gideon Eden have done a superb job of combining the history, concepts, mathematics, and procedural possibilities of the impedance method into one very readable volume. There is a great deal of information in this book. Approaching it from a background steeped in plate counts, you may often be startled at the conceptual jump needed to appreciate the beauty and potential of the impedance method. Even I, who have followed developments in the field for years, learned much from the manuscript. The Edens' influence may not ever rival Koch's, but will undoubtedly be felt for years as impedance microbiology permeates the analytical laboratory.

A.N. SHARPE

Table of Contents

PREFACE   v
ACKNOWLEDGEMENT   vii
FOREWORD   ix
TABLE OF CONTENTS   xi

CHAPTER 1. Introduction-Rapid Automated Methods   1
    1.1 Automation of routine procedures   1
    1.2 Instrumentation for identification and
         susceptibility tests   2
    1.3 Instruments for quantitation of microorganisms   2
    1.4 Comparison between quantitative instrumental
         results and standard methods   5
    1.5 New parameters for quantitative microbiology   5

CHAPTER 2. What is Impedance Microbiology?   7
    2.1 History of impedance measurements in
         microbiology   7
    2.2 Impedance conductance and capacitance   9
    2.3 How bacterial growth changes the impedance   11
    2.4 Factors affecting impedance detection times   18

CHAPTER 3.	Mathematical Analysis of the Impedance Method	21
	3.1 Introduction	21
	3.2 Conductance and capacitance associated with an electrochemical well	21
	3.3 The electro-bacteriological model	23
	3.4 The valid range of the model	28
	3.5 Practical conclusions	30
	3.6 Experimental verification	34
CHAPTER 4.	Considerations in the Development of Impedance Proceedings	41
	4.1 Introduction	41
	4.2 Differences between impedance and traditional methodology	41
	4.3 Curve interpretation	43
	4.4 Quality of curves	46
	4.5 Minimization of generation time differences	52
	4.6 Selecting between impedance, conductance and capacitance	57
	4.7 Effect of pH changes of conductance and capacitance	62
	4.8 The effect of sample preparation	64
	4.9 Neutralization and/or separation of inhibitors and interfering substances	67
	4.10 Sampling errors	70
CHAPTER 5.	Estimation of Microbial Levels	73
	5.1 Introduction	73
	5.2 The calibration curve	74
	5.3 Accuracy of the plate count method	74
	5.4 Accuracy of the impedance method	76
	5.5 Samples required for a calibration curve	77
	5.6 Methods for obtaining samples with above average bioburdens	80

	5.7 How to use the calibration curve for QC	81
	5.8 Food applications	84
	5.9 Clinical applications	89
	5.10 Selective media	90
CHAPTER 6.	Utilization of Impedance for Non-Counting Application	99
	6.1 Introduction	99
	6.2 Shelf-life	100
	6.3 Sterility testing	103
	6.4 Sensitivity to antibiotics	107
	6.5 Characterization and identification	113
	6.6 Microbial kinetics	120
CHAPTER 7.	Instrumentation	123
	7.1 General considerations	123
	7.2 Impedance measuring circuits	125
	7.3 The Bactometer® Microbial Monitoring System M123	128
	7.4 The Malthus Microbiological Growth Analysis	134
APPENDIX A	Statistical Analysis of Scattergrams	141
APPENDIX B	Alphabetical List of Terms	153
REFERENCES		159
INDEX		169

CHAPTER 1
Introduction:
Rapid Automated Methods

The age of automation has touched many areas of science and technology, and it seems as if microbiology is one of the last areas to be affected. This is not because there has been a lack of interest in the field of automation by microbiologists. The proceedings of international symposia (Johnston and Newsom 1976, Tilton 1982) demonstrated a great interest, at the research level at least, in a wide range of methods. The main interest has been in the medical area. However, industrial and food microbiologists seem to be similarly motivated to develop rapid automated methods. This increasing interest is doubtless caused by rising laboratory workloads without parallel rise in budget.

1.1 AUTOMATION OF ROUTINE PROCEDURES

The first area in microbiological laboratories to be affected by automation was that of routine and mundane procedures. In this category we include automated pipettors and dilutors, automated streaking and incubation, colony counters and staining machines. Two systems are worth mentioning in this category; 1. The spiral plater (Spiral System) which has a dispenser that distributes a continuously decreasing volume of liquid on the surface of a rotating petri dish. The dispenser moves from the center toward the edge, resulting in an attenuating Archimedes spiral of colonies. On these dishes a range of 10,000:1 can be read on a single dilution (Gilchrist et al. 1973). 2. The hydrophobic grid membrane filter

(ISO-GRID® HGMF, QA Laboratories) which is essentially a membrane base upon which a hydrophobic grid is applied to divide the membrane into a large number of individual growth compartments. The HGMF count is determined by a most-probable-number calculation (Brodsky et al. 1982). By providing an array of many (e.g. 1600) discrete growth compartments, it permits the organism to be enumerated over a range of nearly 4 log cycles on a single plate. This system retains many of the attractive features of membrane filtration (e.g. concentration of bacteria, removal of interfering substances, resuscitation of injured organisms). However, the need for filtration causes problems with certain samples which contain small particles, fat globules and other interfering substances. Some samples (e.g. dairy products) require an enzyme digestion step prior to filtration.

1.2 INSTRUMENTATION FOR IDENTIFICATION AND SUSCEPTIBILITY TESTS

Several instruments are available for the identification of microorganisms, or for testing their susceptibility to antibiotics. These instruments are mainly geared to the clinical area but may also be used in industrial and food microbiology. Most of these instruments analyze turbidity as a measure of the concentration or types of nutrients needed, or, conversely, the lack of growth in the absence of growth factor or in the presence of inhibitors. The Autobac (Pfizer Diagnostics), the MS-2 (Abbott Diagnostics Products) and AMS (Vitek System, subdivision of McDonnell Douglas Corp) are representative of the instruments available. The AMS System is the most automated instrument involving turbidity. It combines both detection and identification schemes which, in many cases, can bypass the initial requirement for a pure culture.

1.3 INSTRUMENTS FOR QUANTITATION OF MICROORGANISMS

The area where the penetration of instrumentation is the slowest is in estimation of microbial loads. The desire for direct enumeration has inhibited the development of appropriate instrumentation and locked the microbiologist into the petri dish method that was introduced in the 1880's.

A. COUNTING INSTRUMENTS

Some of the instruments available in this area directly count numbers of organisms. An example of such instrument is the Coulter Counter (Coulter Electronics) which uses electronic particle sizing to detect and enumerate microorganisms. Direct microscopic counts can be done automatically using the Foss Bactoscan System (Dickey-John Canada Inc.). This system was mainly designed for the dairy industry. Similarly the direct epifluorescence filter technique (DEFT) can also be used. With the DEFT method the microorganisms are recovered from the sample on the surface of a membrane filter. The retained cells are stained with fluorescent stain, and counted by an epifluorescence microscope (Pettipher et al. 1980).

B. NON-COUNTING METHODS

The most promising rapid and automated methods for the estimation of microbial loads are those which do not involve counting. In this category we include very sensitive analytical instruments that can detect microbial components or metabolites. The impedance method belongs to this category. Since this book deals with the impedance method we will not discuss it here, but rather concentrate on other systems based on different principles.

i. <u>Microcalorimetry</u>

The minute amounts of heat produced by growing cultures can be detected by a sensitive calorimeter. Flow microcalorimetry was used to detect heat production by bacteria which commonly cause urinary infections. This method could detect the presence of bacteria at concentrations of $10^3 - 10^5$ organisms/ml, depending upon the bacterial strain (Beezer et al. 1978). The profile of heat production versus time, or thermogram of microorganisms has been shown to be sensitive to growth conditions. The change in response due to small changes in medium composition could be a serious limitation of microcalorimetry. Although this method was suggested over 10 years ago, this methodology has not been commercialized for quantitative tests of microorganisms.

ii. Radiometry

This method detects radioactive carbon dioxide produced by bacterial metabolism of a radio-labeled source of carbon incorporated into a culture medium. The time required to reliably detect radioactive CO_2, is inversely related to the initial number of organisms in the sample. The Bactec-460 (Johnston-Laboratories) tests the head space gases above inoculated, labeled media at a rate of 60 samples/hr. This instrument is widely used for clinical samples (mainly blood cultures); however, its use in other areas has not been widely reported.

iii. Bioluminescence

Microorganisms, like all living matters, contain ATP. When the luciferin-luciferase enzyme-substrate system comes into contact with ATP, light is emitted. Since the amount of light is proportional to the amount of ATP present in the system, the assay can be used to determine the number of organisms present. Non-microbial ATP is present in many systems. This must be destroyed before the microbial ATP assay begins. The sensitivity of the method depends on the quality of its releasing agent and the quality of the photometer. Lumac (Medical Product Division/3M) reported on a complete ATP System, including all needed reagents. The procedure, however, requires considerable sample preparation before the actual ATP measurement can take place. New instruments have been developed in which the reagents are added automatically. An example of such an instrument is the Picolite™ Luminometer (United Technologies Packard) in which 48 samples can be loaded into a sample changer, where the reagents are automatically injected.

1.4 COMPARISON BETWEEN QUANTITATIVE INSTRUMENTAL RESULTS AND STANDARD METHODS

Every new quantitative technique is eventually compared with existing standard methods. In most cases this entails regression analysis to show appropriate correlation and calculation of confidence limits for such comparisons. There are methods for which such comparison is straightforward. The comparison of direct microscopic count to plate counts is such an example, since both techniques estimate similar quantities, i.e. clumps of organisms which might be considered as colony forming units.

The comparison becomes more difficult with non-counting methods, since the different procedures are based on different assumptions. For example, in the comparison between ATP and plate count we could agree that ATP is a more basic unit than CFU since CFU can only be defined as an entity that gives rise to a colony (Wood and Gibbs 1982). However, since the ratio ATP/CFU is not necessarily constant, it is quite possible that new methods will bear a close relationship to the standard method only under very particular circumstances.

1.5 NEW PARAMETERS FOR QUANTITATIVE MICROBIOLOGY

In reviewing the new microbiological technologies one must raise the question of whether the current counting methods are the best, or even appropriate for estimating the microbial quality of products. It might very well be that a new method can better estimate shelf-life, safety, and acceptability of food products than the standard counting methods. Sharpe (1980) argues that microbiologists are obsessed with the counting of "clusters of bacteria which can grow on agar". This obsession has greatly retarded the development of more realistic methods and, particularly, instrumental methods for assessing the quality of products. Sharpe (1979) further indicates that the ability of food to cause illness, or otherwise become unwholesome, depends on five factors; (i) number of organisms; (ii) their rate of multiplication in the food; (iii) the contribution of each microbial cell to the unwholesomeness of the food; (iv) inhibition or stimulation of (ii) and (iii) by other organisms; and (v) the consumer response thresholds to the levels of the various parameters. Counting microorganisms provides information on only the first item of the above list. Methods measuring levels of some physical, chemical, biochemical characteristics of the measured organism can provide information on factors (i) to (iv). Therefore, it is probable that these new methods, such as impedance, are a better direct measurement of the quality of foods than the standard plate count method.

CHAPTER 2
What is Impedance Microbiology?

2.1 HISTORY OF IMPEDANCE MEASUREMENTS IN MICROBIOLOGY

Although impedance monitoring, as performed by modern instruments, is a relatively new procedure, the observation of impedance variations due to microbial metabolism was reported to the British Medical Association as early as 1898 by Stewart (1899). He showed that the conductivity, measured daily, of defibrinated blood allowed to putrefy, increased tenfold over a course of 25 days. He postulated that ions were formed by bacterial decomposition of proteins and fats present in blood, and suggested that the rate of bacterial growth could be monitored by electrical means. This work is astonishing when one considers that Arrhenius had published his theory of ionic dissociation only eleven years earlier.

A few years later, Oker-Blom (1912) observed that conductivity measurements reflect collective changes taking place in media which can give a general picture of microbial metabolic processes, and lamented that this valuable technique had not been generally accepted among the major methods of measurement in bacteriology.

In 1926 Parsons and Sturges (1926 a,b) and later Parsons et al. (1929), working for a meat packing firm, reported upon the rather exact relationship between conductivity changes and ammonia production by cultures of <u>Clostridia</u> species grown in various media under anaerobic conditions. The conductivity bridge used had an accuracy of 0.5% and gave conductivity measurements which predicted ammonia production within 5% of the average of the experimentally measured values. Still, only daily measurements were taken and typical measurements were conducted over ten to twenty days. The authors suggested that other metabolic events such as carbohydrate fermentation and the effects of antiseptics upon growing cultures could be measured conductimetrically. A decade later Allison et al. (1938), extended this work and were able to report very good correlations between conductivity variations and changes in pH, carbon dioxide production, ammonia production and numbers of organisms.

A novel study was performed 32 years later by McPhillips and Snow (1958) in Australia. Using a unique toroidal conductivity cell, these investigators were able to follow conductimetrically acid production in milk caused by the growth of Streptococcus lactis. These measurements, of conductivity versus time, made hourly, resulted in curves very similar to the curves of impedance versus time obtained with modern instruments. Their apparatus consisted of a hollow glass toroidal (doughnut-shaped) sample chamber which had primary and secondary coil windings wrapped about opposite sides of the chamber. Conductivity could be measured with an error of less than one part per thousand. The outstanding feature of this apparatus was the absence of metal electrodes in direct contact with the test solution, thus avoiding any electrode polarization effects.

Wheeler and Goldschmidt (1975) investigated an impedance-based method to estimate the number of microorganisms present in clinical urine samples. When electrical measurements were made on washed bacteria immersed in distilled water and utilizing a 10 Hz voltage source, there was a direct relationship between the concentration of the cells and the resultant impedance. A correlation between impedance and final viable count, within the range of 10^3 to 10^9 organisms/ml, was observed within a 5% error. Since growth was not required for this procedure, the results could be determined within 10 to 15 minutes. Schwan (1963), who earlier investigated the effects of biological samples on electrode polarization, explained that cells near the electrodes cast a "shadow" on the latter. The cells are very poor conductors at low frequencies and therefore force the electrical current to bypass them. This effectively reduces the electrode area which the current reaches and the impedance is increased because it is inversely proportional to the total electrode area reached by the current. He noted, however, that those measurements strongly depended on the electrode preparation itself. Since any cell increases the measured impedance, this method cannot distinguish between viable and nonviable organisms.

In the mid-1970's, Ur and Brown (1975) and Cady (1975), working independently, described the use of continuous impedance monitoring as a tool of wide potential use in clinical microbiology. Both investigators used the sample and reference chamber approach, in which both chambers are filled with identical medium, only one of which is inoculated. The sample and reference chambers form opposing arms of a bridge circuit which is initially balanced. Ur's instrument recorded the offset voltage as metabolic changes unbalanced the bridge. Cady's instrument automatically balanced the bridge and recorded the impedance ratio.

2.2 IMPEDANCE CONDUCTANCE AND CAPACITANCE

Prior to any discussion relating to the impedance technique in microbiology it is essential to define the terminology associated with this method and explain its components.

A kink in a garden hose impedes the flow of water, producing a pressure drop and conversion of mechanical energy to heat. Similarly, the flow of electric current (i) always encounters some resistance, resulting in voltage drop (v) and the conversion of electric energy to heat.

An ideal resistance is an energy consuming element obeying Ohm's law:

$v = Ri$

which means that voltage is directly proportional to the current. The proportionality constant R is the value of the resistance, and the units are ohms (Ω):

$$R = \frac{v}{i}$$

For our purposes, it is also convenient to define conductance G as the reciprocal of resistance; that is:

$$G = \frac{1}{R}$$

Conductance is measured in inverse ohms (mhos) or siemens (S) units.

When an electric field is imposed on an electrolyte solution the ions will tend to migrate; the cations move toward the cathode (negative electrode) and the anions toward the anode (positive electrode). This migration of ions constitutes the flow of current in the solution, and each ion carries a fraction of the current proportional to its mobility and concentration.

Capacitance is an element that stores energy in an electric field but does not dissipate it. A capacitor typically consists of two conducting surfaces or plates separated by a dielectric material. The dielectric prevents current flow when the applied voltage is constant (DC), but a time-varying voltage (AC) produces current proportional to the rate of voltage change, namely:

$$i = C \frac{dv}{dt}$$

The proportionality constant C is the capacitance measured in farads (F). Since the farad is a huge quantity, practical capacitors have values more conveniently expressed in microfarads ($1\mu F = 10^{-6}$ F).

When two metal electrodes are immersed in a conductive medium, each electrode-solution interface can be represented by a series combination of a capacitor and a resistor (Warburg 1899, 1901).

Since the bulk solution is characterized by a pure conductive element, the whole system can be represented by a series combination of a pure resistance R and a pure capacitance C (Figure 2.1). If an alternating sinusoidal potential is applied to the system, the resultant current will depend on the impedance Z of the system which is a function of its resistance R, its capacitance C, and the applied frequency f (cycles/second or Hertz):

$$Z = \sqrt{R^2 + (\frac{1}{2\pi fC})^2} = \sqrt{(\frac{1}{G})^2 + (\frac{1}{2\pi fC})^2}$$

therefore, any increase in conductance G and/or capacitance C results in a decrease of the impedance Z and an increase in the current. The AC equivalent of the conductance G is the admittance Y defined by:

$$Y = \frac{1}{Z}$$

and measured in inverse ohms (mhos) units.

As will be explained in the next sections, microbial metabolism usually causes an increase in both conductance and capacitance. Therefore, as microorganisms grow the impedance decreases. It is believed that it is more convenient for microbiologists to follow a signal that increases with growth than one that decreases. Therefore, throughout this book we have chosen to plot impedance^{-1} or Y as a function of time rather than impedance.

2.3 HOW BACTERIAL GROWTH CHANGES THE IMPEDANCE

Let us start with the resistive component, i.e. the conductance. The changes in conductance are associated with changes taking place in the solution or in the bulk electrolyte. As microorganisms metabolize they create new end products in the medium. Generally, uncharged or weakly charged substrates are transformed into highly charged end products, for example proteins are metabolized to amino

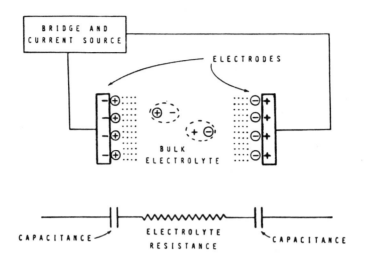

Fig. 2.1 The series resistance and capacitance circuit as a simple model for the effect of electrolyte resistance and double layer capacitance at the surface of the electrodes.

acids, carbohydrates to lactate and lipids to acetate, which increase the conductivity of the solution.

The conversion of one molecule of the nonionized nutrient glucose to two molecules of the ionized metabolite lactic acid by microorganisms increases the conductivity of the growth medium. Further metabolic activity, in converting three molecules of oxygen and one of lactic acid to three molecules of carbonic acid, further increases the conductance since there are now three ion-pairs where there was originally one. In addition, the smaller bicarbonate ion is more mobile and thus a better electrical conductor than the lactate ion. In a similar fashion, a highly ionized molecule such as a protein or nucleic acid will contribute more to the conductance if it is cleaved by enzymatic activity into smaller fragments of equal total charge but of greater mobility. This is, of course, in addition to the ion pairs created by the hydrolysis.

The growth of some organisms such as yeasts, does not result in large changes in conductance. This may be due to the fact that these organisms do not produce strongly ionized metabolites, but rather nonionized metabolites, such as ethanol. In addition, yeasts have the ability to absorb ions from the solution (Suomalainen and Oura 1971). Hence, under certain conditions the conductivity may decrease rather than increase.

Other, more indirect effects, play a role as well. Hydrogen ions (hydrated protons) are nearly seven times more effective as conductors than sodium ions; therefore, one might predict that weakly buffered media would allow a greater conductance change than more strongly buffered media.

Most attention has been given to the effects of microbial growth on the conductance of the medium. Very little attention has been given to its effect on the capacitance due to the electrode polarization. The ions of the electrolyte near the electrode do not experience the same environment as in the bulk solution. Ions are subject to certain forces on the side facing the bulk solution and to different forces on the side facing the electrode. Consequently, the electrode develops a certain electrical field structure and the bulk solution develops another. At the interface region, the molecules and ions form a structure which is a compromise between the structures dictated by both phases. There is a new arrangement of solvent dipoles and charged species. The electroneutrality of this region is broken and the electrolyte side of the interface becomes charged.

Each charge induced on the metal electrode is associated with an equal and opposite charge on the electrolyte side of the interface. The result is a separation of charges across the electrode-electrolyte interface. In total, the interface region is electrically neutral with a potential difference across the interface. The arrangement of charges and oriented dipoles in the interface region of a boundary electrolyte is called the electrical double layer.

Many models have been created to explain the behavior of the electrical double layer at the electrochemical interface. The first was described by Helmholtz (Mott and Watts-Tobin 1961). In this model the first row of solvated ions near the electrode is oriented by the charges on the electrode. The second row is occupied by solvated ions (Figure 2.2). These two layers of charge comprise the double layer and can be described by the capacitor shown in Figure 2.2C. A more complex model is shown in Figure 2.3. In this model it is assumed that the hydrated ions approach the electrode until they touch this layer of water molecules. The ions can also be desolvated on the side facing the electrode and be absorbed by the electrode itself. Both the adsorbed ions and the ions in the compact solvated layer affect the capacitance. The effect of ions in the solution outside the solvated layer is negligible. The capacitance is affected by the dielectric constant of the solution (ε), the area of the double layer (A), and the thickness of the dielectric (d) as follows:

$$C = \frac{\varepsilon}{4\pi} \cdot \frac{A}{d}$$

Microbial growth generating smaller ions can decrease the distance between the capacitor plates. The smaller ions can also increase the effective surface area of the plate by increasing the concentration of ions in close proximity to the electrode. In addition, the newly formed metabolites may have a different dielectric constant.

It was found that pH plays a major role in the polarization capacitance (Firstenberg-Eden and Zindulis 1984). As the pH decreases the capacitance increases, which might be due to hydrogen ions decreasing the distance (d) and increasing the effective area (A).

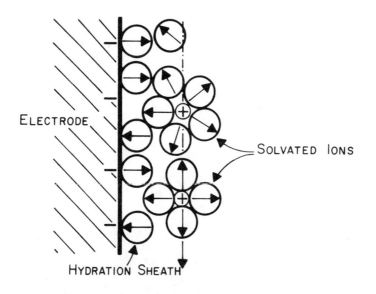
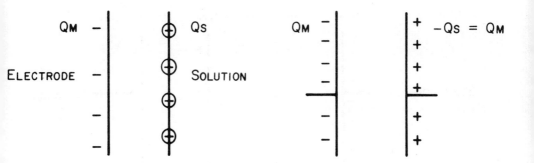

Fig. 2.2 Helmholtz model of double layer and equivalent capacitor.

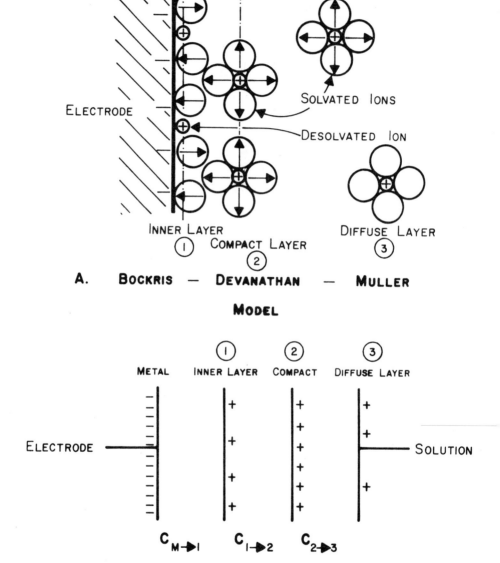

Fig. 2.3 Bockris-Devanathan-Muller model of the double layer and its equivalent capacitor.

So far we have discussed in general terms how microbial growth can change the electrical impedance. The upper graph in Figure 2.4 shows the relationship between the concentration of microorganisms and time, in typical growth curves. The corresponding impedance curves are shown in the lower section of Fig. 2.4. In this example, 10^7/ml organisms are required to produce a detectable acceleration in the impedance curve. The time required to reach this threshold concentration is called the "detection time" and is a function of both initial concentration and the growth kinetics of the organisms in a given medium. After the threshold is passed, the impedance change with time may be proportional to the number of viable cells. The threshold level of microorganisms (10^7/ml in this example) is a function of the type of organism, type of medium, and type of electrodes.

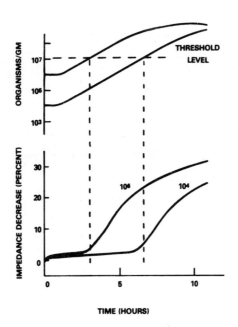

Fig. 2.4 The relationship between bacterial growth curves and impedance curves. (A threshold of 10^7 org/gm is assumed).

2.4 FACTORS AFFECTING IMPEDANCE DETECTION TIME (IDT)

The time at which acceleration can be seen in the impedance curve depends upon the physical conditions of the system (i.e. temperature, media, electrode type) and the microbial characteristics (i.e. microbial concentrations, metabolism and generation times).

A. CONCENTRATION OF MICROORGANISMS

Figure 2.5 shows the impedance curves obtained with naturally contaminated meat samples having various levels of contamination. It shows that the acceleration occurred earlier in samples containing higher concentrations of bacteria than in those with lower levels.

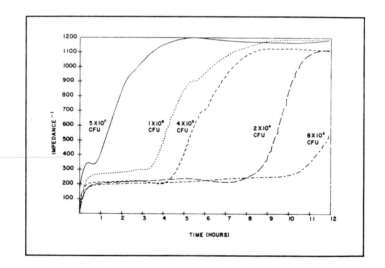

Fig. 2.5 Effect of bacterial contamination level on impedance curves. (From Firstenberg-Eden 1983).

B. GENERATION TIMES

It is important to remember that IDT will correlate with initial concentration only if the generation time of the test population is approximately constant under the experimental conditions. Figure 2.6 shows the effect of different generation times (tg) on the IDT of a certain number of organisms. For example 10^3 organisms with a generation time of 10 minutes will be detected in 2.2 hours (4 log cycles x 3.33 doublings per log x 10/60 hr.). A generation time of 30 minutes will result in an IDT of 6.7 hours for 10^3 organisms/ml while a generation time of 90 minutes will result an IDT of 20 hours.

C. ELECTRODE-TYPE

Since different electrodes provide different sensitivity to conductance or capacitance, the electrochemical perturbation needed before an impedance change is detected will depend on electrode type. Therefore, the IDT of the same organisms in a certain medium will also depend on the type of electrodes, their location, and configuration. Our experiments have shown that electrodes located at the bottom of a bottle resulted in thresholds one log cycle lower than with the same electrodes located at the top of the bottle touching the surface of the solution.

D. CONCENTRATION OF GROWTH MEDIUM

This affects the impedance signal in at least two different ways: (i) At low initial ionic strength (dilute media), a given change in conductance due to microbial metabolism will be relatively large. (ii) At lower concentrations growth media will not have the same levels of nutrients and therefore might result in slower growth.

E. TEMPERATURE

Bacterial generation times are affected by temperature, therefore, changing the temperature will affect the detection time. In addition both components of impedance (capacitance and conductance)

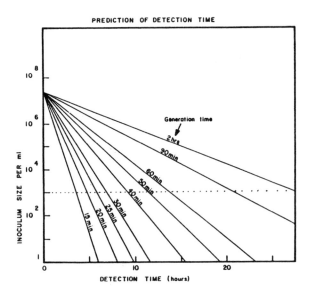

Fig. 2.6 Effect of generation time on detection time.

are temperature dependent. A temperature increase of 1°C will result in an average increase of 0.9% in capacitance and 1.8% in conductance. Therefore if the temperature is not kept constant the impedance patterns might show changes in temperature rather than microbial metabolism.

CHAPTER 3
Mathematical Analysis of the Impedance Method

3.1 INTRODUCTION

The purpose of this chapter is to present the fundamentals of impedance microbiology. We shall establish an "electro-bacteriological" model which will explain the theory behind the methodology. Some basic terms used throughout this book, such as G_{sol}, C_{pol}, threshold, rapidity, etc. will be defined. It is assumed that the reader has some knowledge of mathematics and is familiar with the representation of microbial growth kinetics.

It is important to note that, so far, only the conductance phenomenon is well understood. The capacitance phenomenon is partially understood, in the sense that only qualitative behavior can be characterized. We still lack some basic understanding of the capacitance dynamics during microbial growth. This in no way reduces the utility of measuring capacitance however.

3.2 CONDUCTANCE AND CAPACITANCE ASSOCIATED WITH THE ELECTROCHEMICAL WELL

The electro-bacteriological model described in the next section consists of a pair of electrodes placed into growth medium into which microorganisms have been inoculated. An alternating current of small amplitude (A.C.) is passed through the medium and the resultant impedance is measured.

When a metal electrode is immersed in a conducting fluid, a D.C. boundary potential is established between the electrode and fluid

(Schwan, 1966). If an A.C. current is passed between two electrodes, (Figure 3.1), the D.C. boundary potential becomes modulated by the alternating potential. The arrangement of charged dipoles in the interface region results in a capacitance polarization effect defined as C_{pol}. It has been shown (Warburg, 1899) that the electrode-electrolyte interface can be represented by a series combination of the electrode capacitance C_{pol} and the electrode conductance G_{pol}, the values of which depend on the frequency of measurement. With a purely resistive electrolyte, defined as G_{sol}, a conductivity well consisting of two identical electrodes can be represented by a series combination of conductive and capacitive elements G_T and C_T as illustrated in Figure 3.1. It is the electrode capacitance C_{pol} and the solution conductance G_{sol}, however, which play the major roles in impedance microbiology. Their values may change dramatically as a result of metabolic activities associated with the growth of micro-organisms.

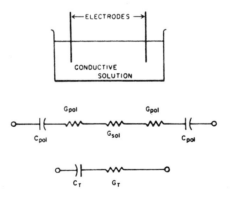

Fig. 3.1 A simplified A.C. model of an electrochemical well. Cpol and Gsol are the polarization capacitance and conductance of each electrode (frequency dependent). Gsol is the conductance of the solution (from Eden and Eden 1984). © 1984 IEEE. *Reprinted, with permission, from IEEE Transaction on Biomedical Engineering, Vol. BME-1, No. 2, pp. 193-198, February 1984.*

3.3 THE ELECTRO-BACTERIOLOGICAL MODEL

As early as 1912 Oker-Blom (1912) observed that conductivity measurements can give a general picture of microbial metabolic processes. However, mathematical representation and practical realization of the value of the technique came slowly. It took about 60 years to establish the impedance method as a practical tool in microbiology (Ur, 1975; Cady, 1975). The model described below is an attempt to explain the experimental observations of conductance measurements (Eden and Eden, 1984).

The observations that during microbial growth the conductance of the medium generally increases are thought to reflect the creation of ion pairs by metabolic activity, such as the conversion of glucose to lactic acid, and the increase in mobility caused by the cleavage of large ions into smaller and more mobile ions. The conductance may increase by 10 to 100 percent during a growth experiment.

In order that the dynamic characteristics of the generated conductance curves may be understood, the basic kinetics of bacterial growth are reviewed. To simplify things, consider a system with a medium into which a single type of microorganism has been inoculated. Figure 3.2 demonstrates the typical change in bacterial concentration as a function of time. It can be seen that there is an initial period during which there appears to be no multiplication, followed by rapid multiplication, and then a leveling off of cell numbers. During the first phase (defined as the "lag phase") the population remains constant whilst individual cells adapt to their new environment. The organisms are metabolizing, but there is no cell division. During the second phase (the "logarithmic phase") the cell numbers double at a constant rate, and during the third phase multiplication ceases, primarily because of exhaustion of nutrients and accumulation of inhibitory substrates.

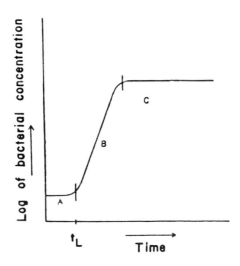

Fig. 3.2 Phases of bacterial growth. A. lag phase; B. logarithmic phase; C. stationary phase. t_L is the time duration of the lag phase.

During the lag phase the bacterial concentration C_B is described by:

$$C_B(t) = C_{BO} \qquad\qquad t \leq t_L \qquad\qquad (3.1)$$

where C_{BO} is the initial bacterial concentration (ml^{-1}) and t_L is the duration of the lag phase (Figure 3.3a).

The usual means of bacterial reproduction during the logarithmic phase is binary fission, i.e. each dividing cell produces two cells. The time required for a cell to divide, or for the population to double, is called the "generation time" (tg). Since the bacterial

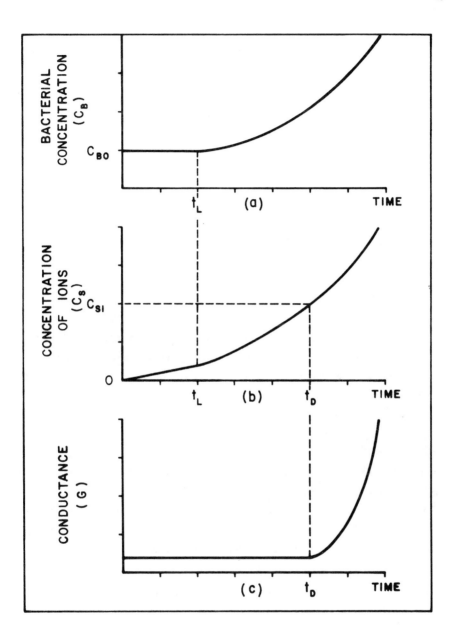

Fig. 3.3 Time variables associated with bacterial growth and metabolism. (a) Bacterial concentration; (b) Ionic concentration due to metabolism; (c) Measured conductance.

concentration at the start of the logarithmic phase is C_{BO}, then after n generations it will be given by:

$$C_B = C_{BO} \cdot 2^n$$

Since the number of generations $n = t/tg$, and since $2^n = e^{n \ln 2}$ then:

$$C_B = C_{BO} e^{t \ln 2 / tg}$$

Since bacterial division is delayed during the lag phase, the instantaneous bacterial concentration C_B during the logarithmic phase (Figure 3.3a) is given by:

$$C_B(t) = C_{BO} e^{(t-t_L) \ln 2 / tg} \qquad t \geq t_L \qquad (3.2)$$

If the microbial metabolism generates ions in the growth medium, and if we assume that each microorganism generates K_B ions per unit time in the solution, then the total concentration of generated ions at any positive time is:

$$C_s(t) = K_B \int_0^t C_B(t) dt \qquad t \geq 0$$

where C_s is the concentration of generated ions, and K_B is defined as a "bacterial activity" coefficient ($\min^{-1}$).

Although during the lag phase the bacteria do not multiply, they are metabolizing and the ionic concentration of the solution will increase. From Equation (3.1) the ionic concentration C_s is:

$$C_s(t) = K_B \int_0^t C_{BO} \, dt = K_B C_{BO} t \qquad t \leq t_L$$

This linear relationship is illustrated between time zero and t_L in Figure 3.3b.

During the logarithmic phase the net ionic concentration is the sum of its value at the end of the lag phase plus the concentration of ions generated whilst cells multiply. Using Equations (3.1) and (3.2) we obtain:

$$C_s(t) = K_B \int_0^{t_L} C_{BO} dt + K_B \int_{t_L}^t C_{BO} e^{(t-t_L)\ln 2/tg} dt$$

$$= K_B C_{BO} t_L + \frac{K_B C_{BO} tg}{\ln 2} [e^{(t-t_L)\ln 2/tg} - 1]$$

$$= K_B C_{BO}(t_L - \frac{tg}{\ln 2}) + \frac{K_B C_{BO} tg}{\ln 2} e^{(t-t_L)\ln 2/tg} \qquad (3.3)$$

The first term in this equation becomes negligible when $t \gg t_L$, therefore, the concentration of ions due to bacterial growth (Figure 3.3b) can be written:

$$C_s(t) = \frac{K_B C_{BO} tg}{\ln 2} e^{(t-t_L)\ln 2/tg}$$

At some instant the concentration of ions C_s generated by bacteria reaches the magnitude of the initial ionic concentration of the solution C_{si} (Figure 3.3b) and measurable increase in conductivity can be detected. This time is called the "detection time" t_D (Figure 3.3c).

$$C_{si} = C_s(t_D) = \frac{K_B C_{BO} tg}{\ln 2} e^{(t_D-t_L)\ln 2/tg}$$

and:

$$t_D = t_L + \frac{tg}{\ln 2} \ln \frac{C_{si} \ln 2}{K_B C_{BO} tg} \tag{3.4}$$

In a later section bacterial concentrations C_{BO}, as determined by plate counts, will be plotted against detection times in a scattergram. The expected relationship can be derived from Equation (3.4) by rearranging its terms:

$$\ln C_{BO} = \ln \frac{C_{si} \ln 2}{K_B tg} - \frac{\ln 2}{tg} (t_D - t_L) \tag{3.5}$$

Since the plate count method usually involves decimal dilutions, it is more convenient to express the initial bacterial concentrations in $\log_{10}$ units (log) rather than in the natural logarithmic base (ln). Conversion of the last equation results in:

$$\log C_{BO} = A' - B'(t_D - t_L)$$

$$\text{where:} \quad A' = \log \frac{C_{si} \ln 2}{K_B tg} \qquad \text{and:} \quad B' = \frac{\log 2}{tg} \tag{3.6}$$

It is, therefore, expected that a scattergram of the logarithms of bacterial concentrations versus detection time will describe a straight line having a negative slope B' and an intercept A'. Both constants are determined by the particular properties of the medium-bacteria combination.

This completes the basic electro-bacteriological model. Its valid range is discussed in the next section. Later, some useful conclusions will be derived from the model which will enable the reader to understand and apply the calibration procedures.

3.4 THE VALID RANGE OF THE MODEL

Every model describing a physical phenomenon has a valid range, i.e. there are theoretical and/or physical limits beyond which it fails to predict the reality. The model discussed in this chapter has a lower limit of applicability resulting from statistical considerations and an upper limit imposed by the model itself.

The lower limit, i.e. the lowest number of microorganisms which still satisfies the linear relationship of Equation (3.6), is, at first glance, boundless. In other words C_{BO} may be 1.0, 0.1, 0.01 organisms/ml, or even lower. This of course makes physical sense only if volumes much larger than one ml are being used. The fact that the model is not restricted by a minimum bacterial concentration, makes the impedance method suitable for any, well defined, sterility test.

Although a lower limit does not exist in this sense, there is an inseparable source of error associated with measurements of low level bacterial contaminations. Detection of very small numbers of organisms in a given volume is subject to sampling errors which are not unique to the impedance method. It is a statistical fact that for lower numbers, simultaneous measurements performed on the same sample may give different results. In the standard method, for example, it has been established (Velz, 1951; Niemala, 1983; Fisher, 1922) that pronounced sampling errors occur if the tested sample contains less than 30 organisms. The impedance method is also subject to this statistical error. Therefore, we can anticipate increased scatter around the regression line as the number of the organisms in the analytical sample approaches the range of the sampling error.

The upper limit is a different story. In developing the model, one finds that the linear relationship was obtained only by neglecting a term in equation (3.3) which assumed detection times much longer than the lag period t_L. For very high bacterial numbers, i.e. samples which detect after very short times, this assumption is invalid, and the relationship is no longer linear but hyperbolic (Figure 3.4). In most cases the existence of an upper limit does not pose any problem, since for most applications detection occurs beyond the lag phase and well within the logarithmic phase. We may, however, anticipate hyperbolic behavior for those applications which for some reason (e.g. presence of growth inhibitors, microbial spores or non-microbial enzymes) are characterized by increased lag periods.

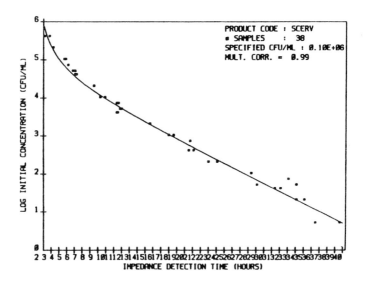

Fig. 3.4   Deviation from the linear model.

3.5   PRACTICAL CONCLUSIONS

The electro-bacteriological model, developed in Section 3.3, establishes a tool for understanding the conductance characteristics of the impedance method. Once this model is verified (Section 3.6) several conclusions may be drawn, relating to practical aspects of the technique, and establishing the course of future developments.

3.5.1 LINEAR RELATIONSHIP

The most obvious conclusion to be drawn from the model is given by Equation (3.6):

$$\log C_{BO} = A' - B'(t_D - t_L)$$

where C_{BO} is the bacterial concentration, t_D is the corresponding detection time, t_L is the duration of the lag phase and A' and B' are constants. The range in which this linear relationship is valid was discussed in section 3.4.

In most cases the user is not interested in the lag phase since detection usually occurs well within the logarithmic phase. Therefore, if one assumes that the lag period t_L is independent of bacterial concentration in a particular application, the scattergram will simply represent the relationship between detection time and concentration. If this reasonable assumption is made, the term t_L in the last equation becomes a constant which can be combined with the other constants. The result is an equation which is convenient for calibration purposes:

$$\log C_{BO} = A - B \cdot t_D \qquad (3.7)$$

where: $A = A' + B' t_L$ and $B = B'$

3.5.2 EVALUATION OF GENERATION TIMES

It follows from Equation (3.6) that the generation time of a specific microorganism growing in a specific environment (medium-temperature combination) can be accurately evaluated by generating the scattergram of $\log C_{BO}$ versus detection time (see next section). The slope of the line which best fits the scattergram is the quantity B of Equation (3.7). The generation time tg is simply $\log 2 / B$.

3.5.3 THE METABOLISM COEFFICIENT

In developing the electro-bacteriological model it was assumed that each microorganism generates K_B ions per minute in the medium. It would be of interest to evaluate this coefficient, which relates to the metabolic activity of the microorganism. Using the scattergram, the ratio K_B/C_{si} can be calculated using the intercept of the line A and the calculated tg. This ratio, which reflects the bacterial ionic contribution relative to the initial ionic concentration of the solution, is called the "relative activity" K_{RB}. The relative activities of different bacteria can therefore be compared if growth is monitored in the same medium which has a fixed value C_{si}. If, however, C_{si} can be approximated (for instance, by evaluating the contributions of conductive elements in a defined medium) then the exact value of K_B can be obtained. The relative activity can be directly derived from the generated scattergram since:

$$\log \frac{K_B}{C_{si}} = \log \frac{B}{\ln 10} - A' \qquad (3.8)$$

it follows that:

$$K_{RB} = \frac{K_B}{C_{si}} = \log^{-1}(\log \frac{B'}{\ln 10} - A') \qquad (3.9)$$

3.5.4 RAPIDITY OF THE IMPEDANCE METHOD

What is the time saving accomplished with the impedance method relative to the standard plate count method? We define the relative rapidity of the impedance method as the ratio of time t_p required by the standard method over impedance detection time t_D for the same sample. The time required for the detection by the plate count method will be usually 24 or 48 hours. Therefore t_p is given by:

$$t_p = K \qquad K = 24 \text{ or } 48$$

The impedance detection time (ignoring the lag period t_L) is (Equation 3.4):

$$t_D = \frac{tg}{\ln 2} \ln \frac{C_{si} \ln 2}{K_B C_{BO} tg}$$

The rapidity ratio defined as the ratio of t_p over t_D is:

$$\text{Rapidity ratio} = \frac{K \ln 2}{tg \ln \frac{C_{si} \ln 2}{K_B C_{BO} tg}} \qquad (3.10)$$

This ratio can also be expressed as a function of the experimental line intercept A' of Equation (3.6):

$$\text{Rapidity Ratio} = \frac{K_B}{A' - \log C_{BO}} \qquad (3.11)$$

As expected the rapidity ratio increases for higher bacterial concentrations. Since more bacteria initially exist in the given volume and each cell is multiplying while generating ions in the solution, a shorter time is required for the ions to exceed the impedance threshold C_{si} than in the standard method which is not sensitive to initial bacterial concentrations.

3.5.5 THE IMPEDANCE THRESHOLD

The impedance threshold is defined as the concentration of organisms at detection, i.e. at the instant at which the impedance curve starts accelerating.

The developed model suggests that the impedance threshold can be derived directly from the scattergram of each bacteria-medium-temperature combination. The logarithm of the bacterial concentration at detection is simply the coefficient A' in Equation (3.6) or, according to Equation (3.7) it can be derived directly from the scattergram at t_L as illustrated in Figure 3.5.

In the following sections it will be demonstrated that the differences between the bacterial thresholds of the impedance and the standard methods is just one of the factors establishing impedance as a rapid method.

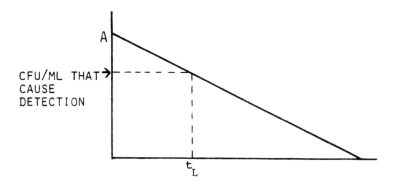

Fig. 3.5 Bacterial threshold as determined by the calibration curve.

3.6 EXPERIMENTAL VERIFICATION

An experiment was carried out to verify the suggested model and to examine the correlation between the impedance and the standard plate count methods.

Fifty five wells containing E. coli bacteria inoculated into Plate Count Broth medium were incubated at 35°C. Their conductances were sampled 10 times per hour. At various times during the experiment aliquots from the wells were plated out in order to determine the concentration of microorganisms. The duration of the lag phase was in the range of 25 - 45 minutes. At detection the concentrations of the microorganisms were $9 \cdot 10^6$/ml. The initial concentrations C_{BO} were also determined by the standard method.

Six conductance patterns are illustrated in Figure 3.6. The initial bacterial concentrations C_{BO} and the corresponding detection times are marked for each pattern. The initial sharp rises in conductances are due to the media warming from room temperature to 35°C.

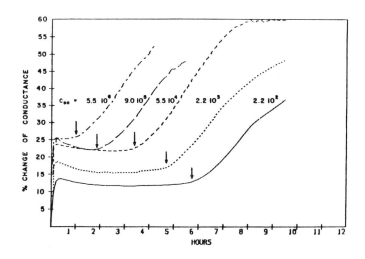

Fig. 3.6 Conductance patterns of samples with different bacterial initial concentrations C_{BO}. The arrow indicates computer determined detection times (from Eden and Eden 1984). © 1984 IEEE. *Reprinted, with permission, from IEEE Transactions on Biomedical Engineering, Vol. BME-1, No. 2, pp. 193-198, February 1984.*

The reproducibility of the impedance curves and detection times for samples having the same initial number of cells is illustrated in Figure 3.7. Differences between the conductance values for each point in time should be ignored: they result from slight differences between the structures of the wells. The variation of the machine detection times for replicates was 0.2 hours on the average and 0.3 hours for the extremes.

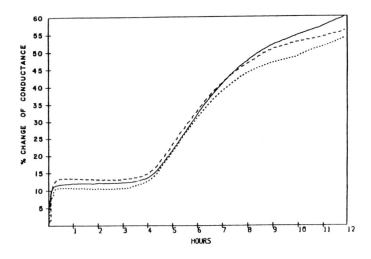

Fig. 3.7 Reproducibility of conductance curves of three samples having identical numbers of microorganisms (from Eden and Eden 1984). © 1984 IEEE. *Reprinted, with permission, from IEEE Transactions on Biomedical Engineering, Vol. BME-1, No. 2, pp. 193-198, February 1984.*

Figure 3.8 is the corresponding scattergram of the experiment. The correlation coefficient between the two methods was 0.98 indicating high agreement between the methods. Using a least-mean-square-error fit, the equation of the linear relationship was found to be:

$$\log C_{BO} = 7.75 - 0.96 t_D$$

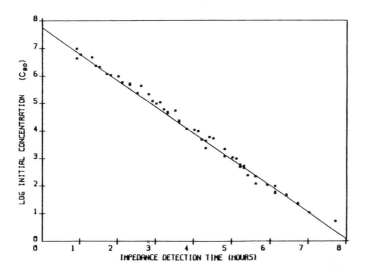

Fig. 3.8 Scattergram of detection times vs. bacterial concentrations determined by the plate count method (from Eden and Eden 1984). © 1984 IEEE. *Reprinted, with permission, from IEEE Transactions on Biomedical Engineering, Vol. BME-1, No. 2, pp. 193-198, February 1984.*

Comparing this relationship with Equations (3.6) and (3.7) and assuming an average lag period t_L = 35 min., the coefficients are given by:

B = B'=0.96
A = 7.75
A' = 7.75-Bt_g=7.75-0.96·0.583=7.19

The generation time for the given combination of bacteria-medium-temperature can now be evaluated using Equations (3.6) and (3.7):

$$tg = \frac{\log 2}{B} = \frac{\log 2}{0.96} = 0.3135 \text{ h} = 18.8 \text{ min}$$

Direct microscopic examination of E. coli cultures (Spector, 1956) indicated generation times of 17 minutes. This examination was performed using the same Plate Count Broth medium but at a higher temperature of 37°C. Since our experiment was performed at 35°C the generation time obtained was somewhat higher.

The metabolism coefficient defined in Equations (3.8) and (3.9) is:

$$\log \frac{K_B}{C_{si}} = \log \frac{B'}{\ln 10} - A' = -7.57$$

or:

$$K_{RB} = 10^{-7.57} = 2.7 \cdot 10^{-7}$$

This ratio indicates the proportional number of ions generated each minute by each microorganism, related to the initial number of the solution ions. It is suggested that this ratio may serve as a tool for the evaluation of metabolic activities of various microorganisms in different combinations of media and temperature.

The bacterial threshold of the impedance method (about 10^7 organisms per ml) reported by Cady (1975) and verified in our experiments was explained in Section 3.5.5. At detection, the concentration of bacteria is given by:

$$C_B(t_D) = C_{BO} e^{\frac{t_D \ln 2}{tg}}$$

where t_D should be corrected to exclude the lag phase in which bacteria are not multiplying. Starting from $C_{BO} = 1$, t_D is given by the ratio $7.75/0.96 = 8.07$ hours. Subtracting the average lag period of 0.583 the modified detection time is 7.49 hours. At that time $C_B = e^{7.49 \ln 2/0.3135} = 1.5 \times 10^7/\text{ml}$ which is similar to the bacterial concentrations observed microscopically.

The rapidity ratio of the impedance method compared to the standard method for the particular medium-bacteria-temperature combination is given by Equation (3.11) and illustrated in Figure 3.9. For very low concentrations of microorganisms the rapidity ratio is mostly determined by the differences between the continuous monitoring of the impedance method as opposed to the constant time intervals allowed by the plate count method. It increases with increasing initial concentration reaching a ratio of 19.4 and 38.7 at $C_{BO} = 10^6$ for plates incubation times of 24 and 48 hours respectively. These calculations suggest that the impedance method can indeed be regarded as a rapid method in which early warnings are provided for highly contaminated samples.

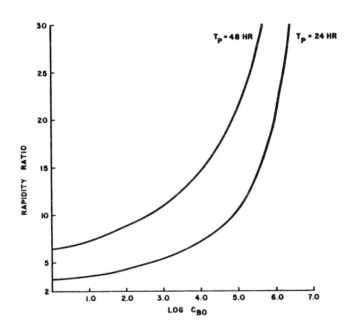

Fig 3.9 Rapidity of the impedance method compared to the standard method as a function of initial bacterial concentration.

CHAPTER 4
Considerations in the Development of Impedance Proceedings

4.1 INTRODUCTION

This chapter should be viewed as an introduction to the next two chapters which deal with various aspects of the utilization of the impedance method. Since the three chapters complement each other it was rather difficult to decide which should be first presented; together they provide an overall view of the development of new procedures and applications of impedance microbiology.

4.2 DIFFERENCES BETWEEN IMPEDANCE AND TRADITIONAL MICROBIOLOGY

There are several fundamental differences between the impedance technique and standard methods for the enumeration of microorganisms. It is important to keep in mind that the impedance technique relies on the measurement of metabolic changes whereas traditional plate count procedures depend on the production of a visible biomass.

Since the impedance technique depends on the rapid detection of metabolic change, factors such as time and temperature become critical parameters in the assay.

A. TIME

When a sample is mixed with growth medium and stored at room temperature or in the refrigerator for an hour the standard plate count may be relatively unaffected. However, since Impedance Detection Time (IDT) is a measure of time, the IDT obtained for a sample stored at room temperature might be shortened by as much as the full storage time, depending on the proximity of ambient temperature to that employed in the impedance method. Changes in IDT for refrigerated samples are even less predictable. Storage at refrigerated temperatures might allow some metabolic activity of certain organisms thereby shortening the lag-time. This will result in shorter IDT for the refrigerated sample than that of unrefrigerated samples. On the other hand some organisms can be lightly injured by refrigeration thereby lengthening their lag time, and their subsequent detection times. These two contradicting effects may result in loss of standardization in the impedance method.

B. TEMPERATURE

When employing the standard plate count a temperature difference of 1 to 2°C during incubation is generally not critical, and the total count at the end of the prescribed 48 hour period will be relatively unaffected. However, a change of 1.0°C will result in an impedance change greater than 1.0%. For this reason temperature control is more critical in incubators employed in the impedance assay than in those for the standard plate count procedure.

Temperature also exerts a significant effect on the generation times of microorganisms present in samples. A change of ±0.5°C in the incubation temperature in the 35.0°C range will result in a difference of 5 to 10 minutes in the generation time for E. coli whereas in the 44.5°C range a similar change might alter the generation time of this organism by up to 30 minutes (Table 4.1). Temperatures employed in the impedance assay must, therefore, be carefully controlled and rigidly standardized.

C. INJURY

Heating, cooling, or freezing the sample can result in sublethal injury to cells which may remain totally unnoticed when the standard plate count procedure is performed following a 48 hour incubation period. This sublethal stress, however, can cause sufficient injury to cells to necessitate a preliminary period of preceding normal replication which lengthens impedance detection times. Therefore, samples from different production steps and their associated types of stress should have separate calibration curves.

TABLE 4.1 Effect of Temperature on the Generation Times of Three E. coli Cultures (meat isolates) Grown in Selective Coliform Medium

Organism	Generation Time (Min) at								
	34°C	35°C	36°C	42°C	43°C	43.5°C	44°C	44.5°C	45°C
E. coli A	23.0	23.0	21.5	20.7	15.1	15.5	18.9	31.5	38.2
E. coli B	22.5	21.0	18.9	14.4	14.8	17.8	24.8	38.1	52.2
E. coli C	26.7	22.4	19.1	17.1	14.9	12.6	21.0	35.4	43.5

4.3 CURVE INTERPRETATION

The success of the impedance technique depends on the quality of curves obtained with the specific sample/medium combination tested. Significant "drift", weak acceleration, and "noisy" curves all tend to obscure visual determination of the onset of acceleration, resulting in inconsistent detections. The requirements for smooth and unambiguous impedance curves are even more critical when IDT is determined by a computer. Several approaches designed to generate "superior" impedance curves are discussed later in Section 4.4.

In the following section we will discuss the significance of specific portions of the impedance curves, present some basic definitions and general guide lines for the evaluation of impedance curves.

A. STABILIZATION TIME

A variable time is required before any impedance curve stabilizes and an acceptably flat baseline is established permitting accurate determination of IDT. The duration of the stabilization period may be affected by a variety of factors including the volume of the tested sample and/or differences between the temperature of the sample and that of the incubator. In the example shown in Figure 4.1 the length of the stabilization period is approximately 1.1 hours.

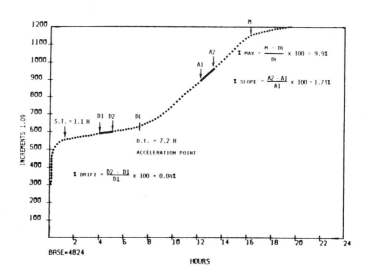

Fig. 4.1 Parameter definitions associated with impedance curves.

B. DRIFT

Drift is another characteristic of the impedance curve and refers to the change occurring in the baseline after the end of the stabilization period and prior to the onset of acceleration. Drift can be quantified as the percentage change per hour in the curve once a stable baseline has been established. In Figure 4.1 the calculated drift is +0.04%/hr for the 1 hour segment occurring between two points (D1 and D2). In curves where the baseline moves downward, the drift value will be negative. When developing a new product protocol, care should be taken to ensure that the generated impedance curves have minimal drift.

C. SLOPE

The "active" segment of the impedance curve starts after the establishment of a stable baseline, where the impedance curve starts accelerating. It continues until the curve starts to decline. A portion of this segment is linear and there we can define the "slope" as the percentage change per hour. Thus in Figure 4.1 a slope value has been calculated for the 1 hour segment between points A1 and A2. In the example shown, the calculated slope value of 1.73%/hr is 40 times greater than the associated drift value (0.04%/hr). This ratio ensures that a computer algorithm will be capable of distinguishing between the active region and the relatively minor changes associated with the drift.

D. MAXIMUM

The "Maximum" value is a property characterizing the _total activity_ in the active region of the impedance curve. This value refers to the total percentage change between the onset of acceleration (IDT) and the maximum level attained during the growth phase. As shown in Figure 4.1, the total percentage change (% Max.) can be determined by examining the segment between points DL and M.

Since the value for % Max. is a reflection of the overall deflection of the impedance curve, and it determines the reliability with which changes may be detected, higher values are more desirable. It is currently suggested that values of approximately 10% or more are required for efficient detection.

E. IMPEDANCE DETECTION TIME

The Impedance Detection Time or IDT is defined as that point (in hours) where the baseline ends and the region of acceleration begins. The Impedance Detection Time for the curve illustrated in Figure 4.1 is 7.2 hours.

Although simple theory leads us to expect that there exists a single detection time for each impedance curve, the observed detection times are subjective measurements which rely heavily on the properties of the inspected impedance curves. In general, those curves characterized by minimal drift and maximum acceleration, produce more distinctive deflections, and minimize differences in calculated IDT's between observers. Such impedance curves also pose fewer problems for automated detection schemes.

4.4 QUALITY OF CURVES

Certain types of media are more useful in impedance testing than others. This may be due to the fact that organisms utilize different metabolic pathways in different media. Since some end products yield a stronger impedance signal than others, the selection of an appropriate medium is crucial. Several examples demonstrating the effect of the growth medium on the quality of impedance curves are described below:

1. Figure 4.2 illustrates impedance curves generated from several coliform media. In this study four commercially available preparations, Brilliant Green Broth (BG), Difco, MacConkey Broth (MAC), Difco, Lauryl Tryptose Broth (LTB), Difco, and EC Broth, Difco, were compared to CM Broth (Bactomatic) which had been specifically designed for impedance testing. The E. coli strain (isolated from meat) grew well in all 5 media. However, the CM Broth resulted in superior impedance curves. When retested with 10 additional coliform strains, similar results were obtained with the CM Broth. As shown in Figure 4.2, curves produced in CM Broth were associated with the steepest slopes

and greatest "maximum" changes. The total amplitude change in CM was more than double that of the other four coliform media tested.

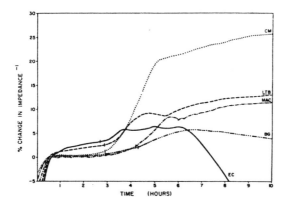

Fig. 4.2 Comparison of impedance curves obtained due to the growth of E. coli M1 in four traditional media: MacConkey (MAC), Brilliant Green Bile (BG), Lauryl Tryptose Broth (LTB), E.C. Medium (EC), and in the medium developed for impedance (CM) (from Firstenberg-Eden and Klein 1983).

EC medium produced markedly inferior curves characterized by "drifty" baselines, an abrupt reversal in the direction of the signal during the growth phase, weak acceleration, and a lower maximum value. Although growth of coliforms in MAC and BG produced acceptably flat baselines, the acceleration "slope" during the growth phase was not sufficiently steep.

2. Figure 4.3 demonstrates another example of the effect of media on the quality of impedance curves. A curve of a meat sample incubated in BHI (upper curve) is compared to that from the same sample incubated in a 1:1 dilution of BHI in peptone diluent

(lower curve). The curve generated in pure BHI is superior since it has a sharper acceleration. This phenomenon could not be attributed to faster growth since equal initial numbers of bacteria yielded identical IDT's in the two media.

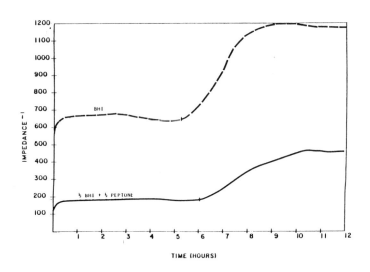

Fig. 4.3 Effect of diluent on the impedance curves. ——Stomached in peptone diluent and 1.0 ml was transferred to 1.0 ml of BHI; --- Stomached in BHI and 1.0 ml was transferred to 1.0 ml of BHI.

3. When grown in BHI many <u>Pseudomonas</u> strains produce impedance curves with a characteristic "hump", these curves show an initial increase, followed by a period showing no detectable change or slight increase, and culminating in a larger and more sustained increase (See Figure 4.4).

This <u>Pseudomonas</u> "hump" is quite reproducible when strains are grown in BHI. The true IDT of such bimodal curves is difficult to assess since two candidates exist at the two

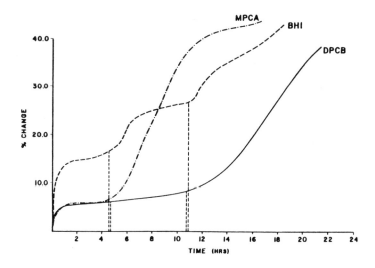

Fig. 4.4 Impedance changes caused by the growth of Pseudomonas in three media: --- Brain Heart Infusion (BHI); —— Double Plate Count Broth (DPCB); and -·-·- Modified Plate Count Agar (MPCA) (from Firstenberg-Eden and Tricarico 1983).

acceleration points. However, when identical Pseudomonas strains were grown in double strength Plate Count Broth, smooth curves without the first hump were obtained. When such strains were cultured on the surface of double strength Plate Count Agar, smooth impedance curves characterized by sharp acceleration coinciding with the beginning of the first hump were obtained. It is interesting to note that this search for a medium that yielded smooth Pseudomonas curves has led us to the discovery that excellent impedance (especially conductance) curves can be obtained with organisms growing on agar surfaces. The hot agar growth medium was first added to the well until it covered the electrodes. The agar was allowed to solidify after which the organisms were added onto the surface.

50

GAS PRODUCTION

Another interesting phenomenon observed in some impedance curves is a characteristic discontinuity at the time when gas associated with carbohydrate fermentation is being produced. Figure 4.5 illustrates the effect of gas production on a typical impedance curve. The net effect of gas formation is the production of random "noise" in the impedance signal. This type of disruption is more pronounced in the conductance signal than in the capacitance signal. Fluctuations in the signal are due to the random release of gas bubbles from the electrode and the subsequent bursting of these bubbles at the surface of the medium. Bubbles exert a direct effect on the electrodes because they can temporarily adhere to the metal surface, thereby decreasing the effective surface area. Once the bubble is released, the surface area of the electrode is restored. When bubbles form in the bulk solution, however, they rapidly float to the top and are released, exerting only a minimal effect on the signal.

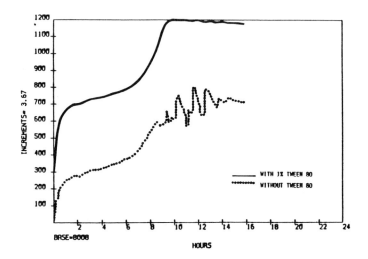

Fig. 4.5 Effect of gas production on impedance curves.

This disruption in the impedance curve can be readily eliminated by the addition of a non-toxic surfactant such as Tween 80 to the medium. Tween reduces the surface tension of the solution, thereby preventing the formation of large bubbles and resulting in an uninterrupted solid impedance curve. Conversely, the disruption due to gas production can be enhanced by the addition of sticky substances (such as grease) to the bottom of the well adjacent to the electrodes. This effect is demonstrated in Figure 4.6 where massive disruption of the signal results from the formation of very large bubbles. The subsequent addition of Tween 80 to the medium effectively minimized this "breakup" in the curve.

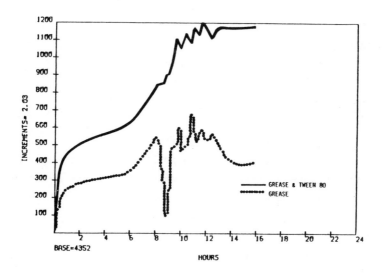

Fig. 4.6 Emphasizing gas production by the addition of grease to the measuring well.

There are situations where the detection of gas production is of diagnostic significance. For example, by definition "coliforms" are organisms capable of fermenting lactose with the subsequent production of gas. Thus when screening samples for coliforms, a medium could be devised that would accentuate the detection of gas released during lactose fermentation. The accuracy of detection would not be compromised since gas production by coliforms generally occurs at a point in the curve after detection has occurred.

On the other hand, some organisms produce large volumes of gas at or near the acceleration region, which interferes with the determination of impedance detection time. In such cases, interference by bubbles produced could be reduced by the addition of surfactants to the medium.

When dealing with a phenomenon such as gas production, the user must first decide whether the effect of gas bubbles on the curve represents a disruptive or beneficial effect. Media formulations can then be adjusted to minimize or potentiate the effect of gas production.

These examples demonstrate how the addition of specific ingredients to the media or changes in media can improve the quality of impedance curves, and the accuracy of the analysis.

4.5 MINIMIZING GENERATION TIME DIFFERENCES

The relationship between generation time and Impedance Detection Time for various concentrations of microorganisms is depicted in Figure 4.7. Although a successful impedance assay does not depend on the presence of identical groups of microorganisms in successive samples, the generation times for all isolates should be similar. Thus, in order to achieve the best possible correlation between standard methods and IDT, the generation times for all organisms should be brought as close together as possible under the experimental conditions employed.

Fortunately, impedance itself serves as a convenient and reproducible method for determining the generation times of microbial populations. Since IDT is a function of the initial concentration of microorganisms inoculated into wells, the higher the dilution, the greater the delay in impedance detection time. By

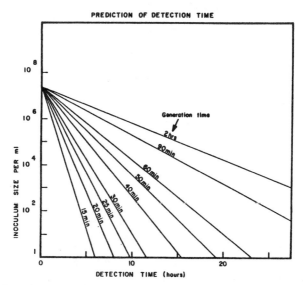

Fig. 4.7 Effect of generation times on detection times of various levels of inoculum.

merely recording the delay in detection times for serial dilutions of the initial sample, the generation time for the culture under the specific conditions can be determined. By using a known dilution factor (e.g. 1:100) it becomes unnecessary to determine the exact concentration of microorganisms in either dilution. The following formula can be employed for detecting the generation times of cultures from delay in IDT for the two serial dilutions.

$$tg = \frac{\Delta IDT \cdot \log 2}{\log n_1 - \log n_2}$$

Since, $\log n_1 - \log n_2 = 2$ and $\log 2 = 0.301$, a constant of 0.15 can be substituted into the previous equation whenever a hundred fold dilution is employed:

$$tg = 0.15 \cdot IDT$$

For example if a delay of 3.0 hours is obtained for a hundred fold dilution, a generation time of 0.45 hour (0.15 x 3.0) or 27 minutes can be inferred.

When differences in generation times are observed, it is important to remember that the farther the initial concentration of organisms is from the detection threshold, the greater will be the impact of minor differences in generation times (see Fig. 4.7). This is because the lower the initial concentration, the greater the number of cell divisions required to reach detectable levels. At the extreme even minor differences in the generation times of populations will be emphasized.

Clearly, when isolates characterized by still wider ranges in generation times (e.g., 30 to 60 minutes) are considered, the resulting spread for samples containing identical initial concentrations of microorganisms may exceed the tolerable range. By first evaluating the major populations present in such mixtures and then determining the appropriate combinations of media, temperature, and concentrations of inhibitors, it is generally possible to bring the generation times of the organisms closer together.

Figure 4.8 compares the generation times (t_g) for two psychrotrophs isolated from a single meat sample as a function of differing incubation temperatures. It can be seen from the results of this study that their optimal growth temperatures are markedly different. The <u>Pseudomonas</u> shows an optimum temperature of approximately 18.0°C while the <u>Coryneform</u> has an optimum between 25.0 and 30.0°C. Differences in their generation times were minimized when incubation was performed at 18.0°C. If such a mixture were to be incubated and tested at 30.0°C, the <u>Pseudomonas</u> would replicate with a generation time roughly 4 times that of the <u>Coryneform</u>. With this difference, even at an initial ratio of 10^5 <u>Pseudomonas</u> to 1 <u>Coryneform</u>, the <u>Coryneform</u> population would reach the instrument's threshold first and thus be the organism detected. Consequently, in most cases only the <u>Coryneform</u> will be detected by the impedance method. While both organisms have the ability to form colonies on plates incubated at 30°C for 72 hours, only the Pseudomonas would be countable on plates since the Coryneform will be diluted to extinction.

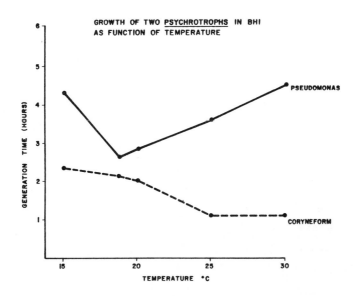

Fig. 4.8 Effect of temperature on the generation times of two psychrotrophs isolated from meat.

The above study clearly indicates that it would be impossible to create a meaningful relationship between IDTs and plate counts with samples containing a mixture of these two organisms if these are incubated at 30.0°C. However, by merely lowering the incubator temperature in the impedance assay to 18.0°C, the IDT's for samples containing pure cultures of either of these two isolates will correlate with plates incubated at 30.0°C. In theory at least, at 18.0°C, 10^5 Pseudomonas will detect in 17 hours and 10^5 Coryneforms will detect in 15 hours permitting the operator to construct a workable correlation.

This example demonstrates how the temperatures routinely used for performing the standard plate count might not be appropriate for use with the impedance method. It should also be stressed that the temperature ultimately selected for use with the impedance assay need not be "optimal" in the traditional sense for any of the individual isolates, but should merely minimize differences in their generation times.

TABLE 4.2 Generation Times of Coliforms in CM at 35°C and 37°C.

Organism	Generation Time (Min)	
	35°C	37°C
C. freundii I	23	39
C. freundii 8090	20	28
E. aerogenes 659	25	42
E. agglomerans	30	54
E. cloacae 23355	49	71
E. cloacae I	25	59
E. cloacae II	31	62
E. cloacae III	20	45
E. coli 25922	23	19
E. coli M2	20	16
K. pneumoniae 609	25	36
K. pneumoniae I	25	36
K. pneumoniae II	23	32
Average	26	41
Standard Deviation	8	16

A similar situation was encountered during the search for a medium for the selective recovery of total coliforms. Some of the standard coliform tests are performed at 37.0°C since incubation at this temperature effectively minimizes the appearance of non-enteric types of bacteria on the plates. When this temperature was employed in an impedance assay, however, fecal coliforms such as E. coli exhibited a very short generation time (16 minutes) while non-fecal coliforms were associated with relatively long generation times (i.e. >1 hour). It was, therefore, necessary to select a temperature where the differences between generation times for all coliforms would be minimized. When the temperature was lowered to 35.0°C, the generation times for all coliforms tested (except 1 species of Enterobacter) fell within the narrow range of 20 - 30 minutes (Table 4.2). A 35.0°C temperature was, therefore, selected as the appropriate incubation temperature for the selective detection of total coliforms using the impedance method.

4.6 SELECTING BETWEEN IMPEDANCE, CONDUCTANCE AND CAPACITANCE

Impedance is a complex entity composed of a resistive component and a reactive component (See Chapter 2). The impedance (Z) can be expressed (as shown previously) in terms of conductance (G) and capacitance (C) as follows:

$$Z = \sqrt{(\frac{1}{G})^2 + (\frac{1}{2\pi fC})^2}$$

where f is the frequency in Hz. Both capacitance and conductance are frequency dependent when measured in an electrochemical well (Schwan 1963). At low frequencies the impedance is predominantly affected by the capacitance, while at high frequencies it is mostly affected by the conductance.

Richards et al. (1978) stated that the capacitance of the electrode polarization is relatively insensitive to the changes accompanying bacterial growth and also is subject to random fluctuations which are of the same magnitude as those ascribed to the growth of bacteria. Their work was done however, with one organism (E. coli) in one medium (PPLO) using platinum electrodes and a driving frequency of 10 kHz. Newer research (Hause et al. 1980 and Firstenberg-Eden and Zindulis 1984) revealed that under different conditions, i.e. stainless steel electrodes and lower driving frequency, capacitance might be very important in microbial monitoring.

Hause et al. (1980) investigated the effect of frequency on an imaginary impedance component (capacitance) and a real impedance component (conductance). The increased response of impedance measurements at low frequencies (100 - 1000 Hz) was a general finding independent of bacterial species, electrode material or electrode configuration. At 100 Hz the impedance signal was almost purely capacitance while at 10,000 Hz it was almost entirely conductance. The changes observed in the capacitance component were always greater than those observed for the conductance components and, in many cases, three times greater. Electrode configuration appeared to have an effect on the relative importance of conductance or capacitance. Electrodes with small area or compact configuration emphasize the capacitance component of impedance, while a series of four parallel electrodes emphasize the conductance component (Hause et al. 1980).

Some instruments allow the operator to separately monitor either the C, G, or Z signal. This permits the operator to initially evaluate all three signals and to select the appropriate signal for monitoring impedance change in a given product/medium combination. Although all of the potential product/medium combinations have not yet been evaluated, some preliminary guidelines for choosing the correct signal (at a frequency of Ca. 1000 Hz) are listed below:

When weakly conductive media (e.g., Nutrient Broth or Plate Count Broth) are employed, bacterial metabolism results in clearly detectable changes in the conductance (G) component which are associated with the accumulation of ionized metabolic end products in the solution. In such cases, measurement of the G signal alone is generally sufficient to detect bacterial metabolism. Although the C signal may also be useful in conjunction with low conductivity media, it is generally characterized by excessive drift and no distinctive acceleration point.

Figure 4.9 depicts changes in the C and G signals caused by the growth of a Pseudomonas species on an agar surface (double strength Plate Count Agar). In such cases the user should monitor only the G signal.

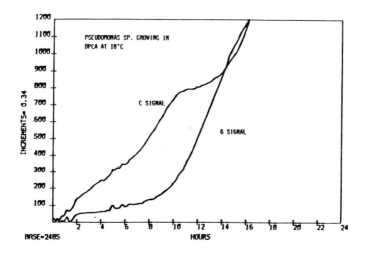

Fig. 4.9 Comparison of the conductance (G) and capacitance (C) signals generated by Pseudomonas grown on the surface of Double Plate Count Agar (DPCA).

A distinctly different situation was observed when yeasts were monitored in carbon base medium with ammonium sulfate (CBAS) (Figure 4.10). Although yeasts produced marked changes in the C signal, the changes in G were minimal. In the example shown, a 20.0% change in the C signal due to yeast metabolism was accompanied by a change of only 2.0% in the G signal. Furthermore, when the G signal was used

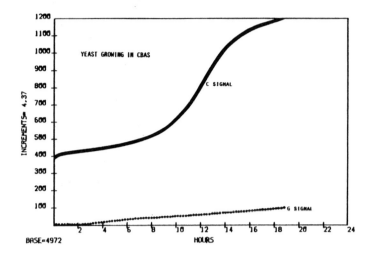

Fig. 4.10 Comparison of conductance (G) and capacitance (C) signals generated by a yeast isolated from orange juice.

to monitor the growth of yeasts, the direction of the conductance curves was unpredictable. While some yeasts resulted in an increase in G, others were associated with a decrease (Figure 4.11). In such cases a mixture of yeasts could potentially produce G signals that would effectively cancel each other. The small change in G obtained with yeasts might be due to the fact that the yeasts do not usually produce strongly ionized metabolites, or may also be due to the ability of yeasts to absorb ions from the solution (Soumalainen and Oura 1971). Removal of ions might have caused the observed decrease in G for C. utilis in Orange Serum Broth (Figure 4.11).

The growth of E. coli in BHI (Figure 4.12) is an excellent example of intelligent use of the impedance signal. Although the overall excursion of the C curves was greater than that of the G curve, the G signal produced a flatter baseline. Furthermore, summation of the two signals may eliminate "glitches" found in some

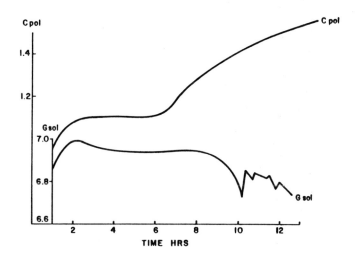

Fig. 4.11 The effect of growth of C. utilis in Orange Juice Serum Broth on capacitance (Cpol) and conductance (Gsol). Cpol in µF and Gsol in mS (from Firstenberg-Eden 1984).

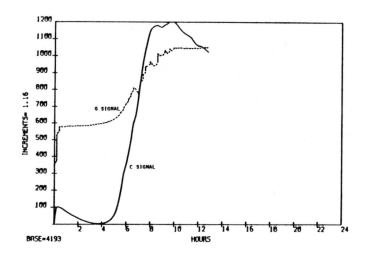

Fig. 4.12 The effect of growth of E. coli in BHI on capacitance (Cpol) and conductance (G) signals.

C curves. In cases where C and G signal appear to be "mutually supportive", it is recommended that the impedance signal (Z) be used.

4.7 EFFECT OF pH CHANGES ON CONDUCTANCE AND CAPACITANCE

When microbial cultures are allowed to progress to their stationary phase, their metabolism typically results in a change of 2-2.5 pH units. The pH changes accompanying microbial growth can be expected to affect both capacitance and conductance.

We determined the contribution of these pH changes to the change of C and G. The relationship between changes in pH and capacitance of three media due to the addition of three organic acids is shown in Figure 4.13, while this effect on the solution conductance is shown in Figure 4.14.

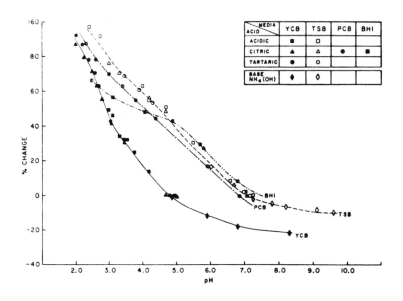

Fig. 4.13 The effect of pH on capacitance (Cpol, μF) (from Firstenberg-Eden and Zindulis 1984).

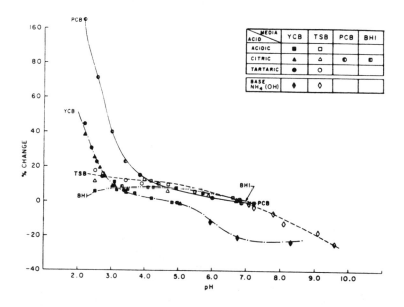

Fig 4.14 The effect of pH on conductance (Gsol, mS) (from Firstenberg-Eden and Zindulis 1984).

A drop of two units in pH resulted in ca. 40% increase in C in all media tested with three types of acids. The addition of base resulted in smaller changes in C (<10%). The fall in pH accounts for 20 - 40% of the total increase in C resulting from bacterial growth (Firstenberg-Eden and Zindulis, 1984) and 20 - 70% of the increase resulting from growth of yeast. Thus pH appears to be an important contributor to capacitance change. This suggestion is supported by the results of an experiment in which the yeast C. utilis was allowed to grow in Yeast Carbon base medium (YCB) with and without added phosphate buffer. In the unbuffered medium a large capacitance change was observed due to the yeast growth (Figure 4.15), while the signal was suppressed in the buffered medium. The increase in G due to the drop of 2 pH units was less than 14%. The fall in pH accounted for 15-35% of the total fall in G resulting from bacterial growth. Therefore, the changes in pH were not the main cause for the change in G in any of the media tested.

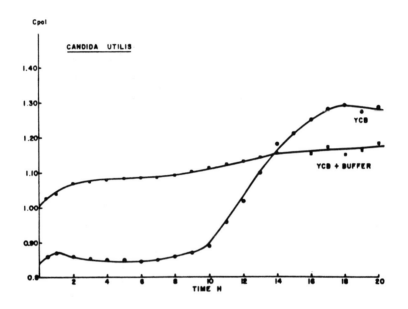

Fig. 4.15 The effect of the addition of buffer to Yeast Carbon Base (YCB) on the percent change in capacitance (Cpol, µF) due to the growth of C. utilis (from Firstenberg-Eden and Zindulis 1984).

4.8 THE EFFECT OF SAMPLE PREPARATION AND STORAGE CONDITIONS

The role of both these factors has been investigated for raw meat samples. A detailed analysis of data obtained by monitoring raw meat samples under a variety of experimental conditions has recently been published (Firstenberg-Eden 1983). It appears that both factors are critical to the quality of the impedance curves.

A. THE EFFECT OF DILUENTS

The diluent in which meat is stomached may affect both the quality of the impedance curve and the impedance detection time. Samples stomached directly into growth media give superior signals and earlier detections compared to those stomached into diluent and subsequently mixed with growth medium (Firstenberg-Eden 1983). This one step procedure also eliminates one of the dilution steps in sample preparation and is, therefore, less labor intensive in the laboratory.

B. STORAGE OF SAMPLES

Storage of diluted samples in the refrigerator prior to testing prolonged the stabilization period. This could adversely affect determination of detection times in some samples. For example, the impedance detection times of refrigerated samples (in peptone diluent or growth medium) were shortened unpredictably when compared with samples placed immediately into the instrument. Preliminary data suggests that storage of raw meat for two hours in the refrigerator (7.0°C) is equivalent to 30 minutes of storage at 30.0°C.

When the diluted samples were stored at the same temperature as the impedance assay (30.0°C), the stored and unstored samples produced accelerations which occurred almost simultaneously (Figure 4.16). However, since the stored samples were added to the instrument 1.7 hours later, the actual detection times obtained on the instrument were shorter (4.8 and 3.1 hours for the stored and unstored sample, respectively).

Storage of meat samples (in peptone diluent) at 30°C prior to loading resulted in wavy baselines which may have been caused by leakage of enzymes and/or ions from the meat cells into the solution. Storage at 30°C in the impedance growth medium, however, resulted in exceptionally smooth impedance curves. Here the IDT of the stored samples was shortened by the exact duration of storage (Figure 4.16).

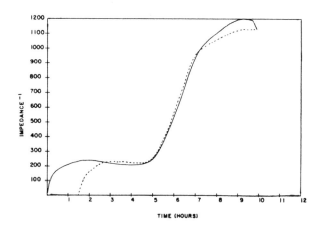

Fig. 4.16 Effect of holding meat samples in BHI at 30°C for 1.7 hours on the impedance curve. Broken line refers to sample held in BHI. Solid line refers to sample monitored immediately (from Firstenberg-Eden 1983).

These studies suggest that, once samples have been diluted, any period of storage prior to monitoring of the impedance samples should be taken into account.

C. EFFECT OF INJURY

Injured cells require time to repair before replication. For this reason, IDT's obtained for samples containing injured cells could be longer than those observed for samples where cells are uninjured.

Impedance detection times for frozen meat samples were invariably longer than those for identical non-frozen samples. There are two possible explanations for these observations. On one hand, the two populations might have been identical, but injured cells in the frozen samples required more time to repair before replication, thereby delaying the impedance detection. On the other hand, freezing might select more cold-resistant populations, characterized by slower growth rate and corresponding impedance detection times.

Regardless of the mechanism responsible, this unavoidable lengthening of IDT for frozen samples indicates that in such situations separate calibration curves will be required for frozen and non-frozen samples.

4.9 NEUTRALIZATION AND/OR SEPARATION OF INHIBITORS AND INTERFERING SUBSTANCES

Certain food products contain natural or added preservatives which may prolong the lag phase and inhibit the growth and detection of microorganisms by impedance. Whenever successive decimal dilutions of a same sample detect at approximately the same time, or when the more dilute sample detects earlier, the presence of an inhibitor can be suspected. It should be emphasized that certain concentrations of inhibitors occurring in a product might not decrease the number of colonies appearing at 48 hours on plates yet affect the IDTs. Since the impedance method depends on the kinetics of microbial growth in broth, levels of inhibitors that might not adversely affect the standard plate count may affect the impedance assay.

A. INTERFERING SUBSTANCES

Certain ions can interfere with the impedance signal by causing excessive drift (in either direction), or by actually reacting with medium components to produce artifacts that mimic microbial growth. To date only a few of these ions have been specifically identified. Chloride ion (Cl^-) for example, causes excessive drift and at high concentration may actually destroy the oxide layer on the electrode surfaces.

Ions that are highly active chemically may undergo spontaneous changes in the medium. Addition of bisulfite or metabisulfite to growth medium results in this type of phenomenon. Depending on pH the bisulfite ion (HSO_3^-) can be converted into either molecular

SO_2 or to sulfite ion ($SO_3^=$) as shown in Figure 4.17. At a pH below 4.0 it will be converted into SO_2 whilst above 6.0, $SO_3^=$ ion will be produced. The SO_2 produced can react with many organic compounds in aqueous solution to form end products referred to as "sulfite addition compounds". Reactions also take place with aldehydes, ketones, sugars, organic acids, thiol groups, enzymes, vitamins, amino acids, and lipids. If a product containing some bisulfite ion is added to a neutral medium, $SO_3^=$ ions could be formed causing a change in conductivity. When the microorganisms present begin to replicate, the pH drops and the $SO_3^=$ ion undergoes additional changes.

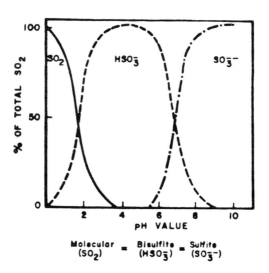

Fig. 4.17 Percent distribution of sulfite, bisulfite and molecular SO_2 as a function of pH in aqueous solution (adapted from Hammond and Carr 1976).

Chemical reactions can also mimic the changes associated with microbial growth. Figure 4.18 demonstrates how the addition of Na_2SO_3 to sterile CBAS broth produces an acceleration which might be mistaken for a microbial response.

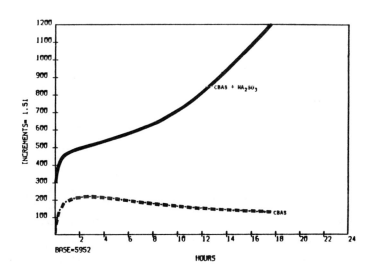

Fig. 4.18 Effect of the addition of Na_2SO_3 to sterile yeast medium (CBAS).

B. METHODS FOR NEUTRALIZING THE EFFECTS OF INHIBITORY SUBSTANCES

The effects of inhibitors or interfering ions in a product can be minimized through dilution or separation procedures. Dilution is acceptable only when sufficiently large numbers of organisms are present in the initial sample. Thus, if organisms are initially present at concentrations of only 10^2 to 10^3 per gram or ml of sample, care must be taken not to dilute them to extinction. When the dilution procedure produces initial concentrations of microorganisms below the tolerable level (i.e., <30/well), a preincubation step may be required.

As an alternative to dilution, organisms can be separated and removed from the product by several techniques including centrifugation, filtration, or the use of resins (Wood and Gibbs 1982). In addition to removing inhibitors, these procedures simultaneously concentrate the organisms. The result is improved signals and earlier detection times. For example, by filtering 20 ml of fruit juices (containing SO_2) through an appropriate filter, not only can we minimize the interference of SO_2 but also achieve a 20 fold concentration of the yeast. This will result in an IDT shorter by 5-7 hours as compared to a 1:10 dilution of product in CBAS.

4.10 SAMPLING ERRORS

A sample can be expected to yield meaningful information only if it is collected in such a manner as to protect it against microbial contamination or from changes in microbial populations between collection and analysis, and provided it actually represents the bulk of material being sampled. After all precautions have been taken to ensure the choosing and uniformity of samples, for example by thorough mixing, blending, or shaking, it must be assumed that: a) the sample is representative of the entire lot of product; b) the subsample used for analysis is representative of the sample; and c) the weighing or measuring of the subsample, diluents and aliquots is accurate.

Various sampling plans have been described in the literature to assure the selection of samples that will yield meaningful results (ICMF 1978b, Kilsby and Pugh 1981). All these plans apply to the impedance method as well as to the plate count method. The main sampling error that this section is dealing with is that obtained after the blending of sample aliquots with diluent.

Any sampling errors due to different aliquots are enhanced when extremely low numbers of organisms are present in the original sample. This is because bacteria contained in such samples follow a "Poisson's distribution". With this type of distribution the standard deviation (SD) is approximately equal to the square root of the mean. Thus if a sample contains 100 organisms/ml the standard deviation should be 10 organisms and approximately about 95.0% of the samples will contain between 80 and 120 organisms/ml. If the mean bacterial level is only 4 organisms/ml the standard deviation is 2 organisms and the 95.0% confidence limit will be between 0 and 8 organisms/ml. Furthermore, in practice, the observed variation between multiple aliquots is usually greater than one would expect from a strict Poisson's distribution. This is because additional variations can occur due to errors in dilution and in the measurement of subsample.

In order to minimize the aliquot sampling error of the impedance assay it is necessary to introduce a minimum of 30 organisms into each well. Lower concentrations of organisms (i.e., <30/well) might result in scattered and unreliable detections. As the initial concentration of organisms/ml in the well increases, the scatter decreases. At very high initial concentrations ($>10^7$/ml), scatter may increase again because the instrument detects microbial by products before the baseline has stabilized.

Precision is further improved for samples containing high concentrations of microorganisms as a result of smaller number of generations occurring before detection. With lower deviations, differences in generation times (tg) became less noticeable and scattergrams of better quality are obtained.

Impedance experiments with coliforms showed that the standard deviation for multiple wells is inversely related to the initial concentration of cells present. The standard deviation was Ca 0.25 hours for samples containing greater than 10^5/ml, 0.3 to 0.4 hour for the range of 10^4 to 5×10^2/ml, and 0.4 to 0.7 hour for the range of 5×10^2 to 5×10^1/ml. At initial concentrations of less than 10 organisms/ml, the standard deviation of multiple wells can be as great as 4.0 hours (Firstenberg-Eden and Klein 1983).

PREINCUBATION

When samples containing low numbers of microorganisms are encountered, sufficiently large volumes of sample suspension are used, as to contain large numbers of organisms. These suspensions are preincubated, thereby allowing the organisms to multiply to concentrations yielding a smaller random error in the aliquots that are subsequently transferred to wells. The preincubation period should be chosen so as to permit the microbial population to reach a level such that at least 30 microorganisms will be introduced into each well. These levels are similar to those suggested for use in conjunction with the standard plate count procedure.

Storage experiments suggest that when sample suspensions are prepared in the medium and held at the temperature of the impedance assay, the preincubation time and IDT are simply additive. Under these circumstances the preincubation time can conveniently be added to the standard IDT to provide a "total impedance detection time".

CHAPTER 5
Estimation of Microbial Levels

5.1 INTRODUCTION

One can envision an unlimited number of potential applications of impedance measurements in industrial food and clinical microbiology. Only a few of these applications have been fully developed to date, others are partially developed, while some are still purely speculative.

The most common application of impedance is to determine if a given sample contains more or fewer than a predetermined concentration of organisms. This approach was used for food products such as meats (Firstenberg-Eden 1983), raw milk (Firstenberg-Eden and Tricarico 1983), frozen vegetables (Hardy et al. 1977), fish (Gibson and Ogden 1980), and in the clinical laboratory for bacteriurea screening (Lamb et al. 1976, Zafari and Martin 1977, and Throm et al. 1977).

A similar screening method was used in conjunction with selective media to detect levels of a variety of groups of organisms of importance in food and industrial microbiology. The groups reported include coliforms (Firstenberg-Eden and Klein 1983), yeasts (Zindulis 1984) and Staphylococci (Rantama 1983).

The general approach used in developing procedures for impedance was discussed in the previous chapter. It is assumed that the reader is familiar with its content. In the following two chapters specific applications will be discussed. The combination of specific applications (Chapters 5 & 6) with the approaches outlined in the previous chapter will enable the reader to develop procedures for his own application.

5.2 THE CALIBRATION CURVE

Prior to adapting an impedimetric method to estimate numbers of bacteria a calibration curve should be generated. This curve defines the relationship between impedance and the standard method, usually the standard plate count (SPC) method. The statistical considerations involved in the construction of a calibration curve are given in Appendix A. In an ideal situation i.e., no error in either method, the plot of log CFU/ml vs. impedance detection time (IDT) is a straight line with no scatter along the line. Such data has a correlation coefficient of -1.00. In practice scattergrams similar to that depicted for hypothetical product X will generally be observed (Fig. 5.1). In this scattergram the experimental results show vertical and horizontal scatter. This scatter is due to the accumulative error in both methods (SPC and impedance). The errors associated with each method are discussed in the next two sections.

5.3 ACCURACY OF THE PLATE COUNT METHOD

Because of its historical status as the reference method, many microbiologists tend to regard the results of the standard plate count as absolute. The true relationship between the standard plate count and the actual number of viable microorganisms in the sample must, however, be carefully evaluated in terms of the limitations associated with this method. Several limitations inherent in the standard plate count method are summarized below:
1. Microorganisms naturally exist in clusters or clumps of viable units which tend to resist fractionation into single cells during the blending or stomaching processes used to prepare samples. Therefore, colonies subsequently forming on agar may be actually derived from one, two or even twenty or more cells. The actual degree of separation of clumps depends on the sample

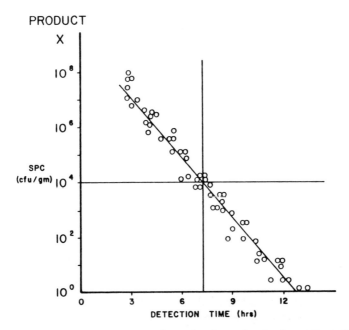

Fig. 5.1 A scattergram for the hypothetical product X. A permissible level of 10^4 CFU/g was assumed.

type, the endogenous microflora, and the preparation process. There is thus only a very vague relationship between numbers of microbes in the sample, potentially able to form colonies and colonies actually counted.

2. To be enumerated a specific microorganism must be capable of forming a visible colony under the actual conditions employed in the assay. This requires that all organisms counted thrive and replicate under the medium and temperature conditions employed, and within the incubation period of the assay. Since there is no universally acceptable medium/temperature/time combination, no plate count method can be expected to include all microorganisms originally present in a sample.

3. Sampling errors inherent in any procedure where different analysts are involved in weighing samples, pipetting dilutions, and enumerating colonies also contribute to the inaccuracy of test results.

The International Commission on Microbiological Specifications for Foods (1978) conducted a comparative and collaborative study of coliforms naturally occurring in various food samples. They determined that 95.0% confidence intervals for a single value reported fell within the range of $\pm$ one log cycle. Analysis of variance indicated that the variation between replicate aliquots was overwhelmingly the most important source of error.

Despite these acknowledged limitations of the standard plate count method, it remains the reference method and any automated procedure proposed as an alternative must correlate with it. If such a comparison is to be made, the standard plate count technique must be performed in a well standardized manner in the hope that this will minimize variation and generate results that are as reproducible as the method to which it is being compared.

It is well known that the precision of the standard plate count method is best when the number of colonies counted is within the range of 30 to 300 per plate. Therefore, in the process of constructing a scattergram it is important to count only plates containing colonies within this range.

5.4 ACCURACY OF THE IMPEDANCE METHOD

There are several error sources in the impedance method. The most important ones are listed below:

1. Large scatter in generation times. It was shown in Chapter 3 that in order to correlate the IDT to microbial numbers the microflora involved (under the experimental conditions) should have similar generation times. If the chosen conditions e.g., temperature, medium, oxygen level are inappropriate, a larger scatter in generation times will result and the correlation of IDT's with actual numbers will be poor.

2. As with the plate count method, the organisms must grow in the chosen medium. In addition, the medium and the sample preparation handling prior to impedance measurement affects the quality of the impedance curves. Poor quality curves result in inaccurate detections.
3. Manipulation errors are smaller with the impedance method since less pipetting and diluting is needed. However, a larger random error may be introduced if low numbers of microbial cells (below 30) are added to the medium in the measuring well.

Since these errors are cumulative special care must be taken to minimize each one in order to obtain an accurate impedance result. Possible approaches to the minimization of these errors were discussed in Chapter 4.

5.5 NUMBER OF SAMPLES REQUIRED FOR A CALIBRATION CURVE

In determining the number of samples required to construct the calibration curve a balance must be struck between the optimal number of data points required for reliable calculation of the curve, and the amount of time and effort required to obtain such data. Generally, as the number of samples analyzed increases, the calibration curve more reliably describes the performance of a particular product in the impedance assay. It has been determined experimentally that 100 samples are generally sufficient for the preparation of a reliable calibration curve having a significant correlation coefficient, and meaningful line parameters.

It is suggested that the 100 samples chosen should have microbial concentrations nicely distributed over a four to five log cycle range. When microbial concentrations are restricted to a narrower range (e.g., one to two log cycles), the regression line is very sensitive to the addition of points in certain areas of the scattergram. In such cases all statistical parameters can change markedly with the addition of data from one or two fresh samples. The regression line probably will not accurately represent the relationship between impedance detection time and standard plate count at higher or lower concentrations forcing the user to extrapolate beyond the limits of his data during routine analysis. "Out of spec" samples are generally of greatest importance to the user and they should, therefore, be included when constructing the calibration curve. It is recommended that at least 20.0% of the samples included in the standard curve have concentrations of microorganisms in excess of the permissible level, and that at least 10.0% have concentrations at least 1 log above the permissible level.

The importance of using samples having contamination levels distributed over four to five log cycles is illustrated by the calibration curve shown in Fig. 5.2. If only those data points which fall within two log cycles (Fig. 5.3) are used, the correlation coefficient changes from 0.95 (Fig. 5.2) to 0.55 (Fig. 5.3). The slope of the curve derived from the partial data is half of that derived from the full data, and the intercept is two log cycles lower. The values obtained for the intercept and the slope for the partial data disagree with the theoretical values (see Chapter 3). Furthermore, on addition of only one point to this partial data (Fig. 5.4) the correlation coefficient changes to 0.72 from 0.55, the slope changes by 40% and the intercept increases by one log cycle.

It must be stressed that the relationship between impedance detection time and log CFU becomes non-linear when very high ($>10^7$/ml) or very low ($<10^2$/ml) concentrations of organisms are present in the wells (see Chapter 3.4). With higher concentrations the threshold level for impedance change may be reached prior to the establishment of a good baseline, making accurate determination of the impedance detection time more difficult. Although the impedance detection times observed for samples with excessively high concentrations of microorganisms are characterized by considerable scatter, they generally fall sufficiently far to the left of the scattergram to alert the operator to the presence of grossly contaminated samples.

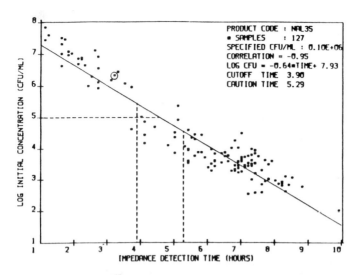

Fig. 5.2 A scattergram for mesophilic bacteria in raw milk containing samples with counts spread over six log cycles.

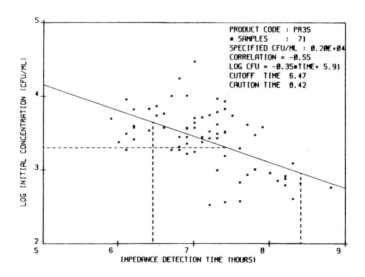

Fig. 5.3 Scattergram obtained from samples having counts between 3×10^2 and 3×10^4 in Fig. 5.2.

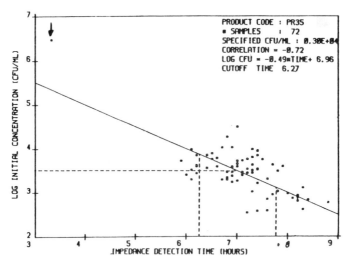

Fig. 5.4 Same data as in Fig. 5.3 with addition of one point (↓) from Fig. 5.2 (circled).

At the other end of the spectrum, samples with low concentrations of microorganisms produce a "tail" in the scattergram. In this area, samples with similar plate counts may exhibit considerable scatter in impedance detection times. This is due to both the uneven distribution of the various types of bacteria in the wells and the imprecision of the standard plate count at these low levels.

The calibration curve should be constructed using only those points occurring within the linear range. However, when samples with extremely low bioburdens are required for completion of the standard curve, special techniques such as preincubation, filtration, or plating of larger aliquots are permissible.

5.6 <u>METHODS FOR OBTAINING SAMPLES WITH ABOVE AVERAGE BIOBURDENS</u>

Naturally contaminated samples usually contain concentrations of microorganisms in a range of one to two log cycles and 95.0% of the samples are generally below spec (defined as the "permissible level"). For this reason a good scattergram can rarely be obtained from only naturally contaminated samples and some form of sample "abuse" must be employed in order to extend the calibration range.

As a general rule the procedure used to abuse samples should closely mimic those conditions which are naturally responsible for higher numbers of organisms in the product. It is important to ensure that the abuse protocol does not produce a "shift" to an unnatural flora as might occur as a result of indiscriminately changing parameters such as water activity (a_w) or temperature. Therefore, the appropriate way to abuse a sample that is normally refrigerated would be to elevate the storage temperature slightly (e.g. no more than 10.0 to 12.0°C). Warming samples too much above the normal storage temperature (e.g., 35.0°C to 37.0°C) will strongly favor development of an unnatural mesophilic microflora. The presence of elevated numbers of this new microflora in abused samples could adversely affect the scattergram.

The organisms in the abused sample must be in the same physiological state as those in the naturally contaminated high count samples. If, for instance, the product is normally frozen abused samples should be produced by thawing the product, allowing the organism to proliferate and then refreezing it before testing. Since both abused and naturally contaminated samples are thawed in the same manner (i.e., immediately before testing) they are more likely to have the same flora and therefore fit on the same scattergram.

5.7 HOW TO USE THE CALIBRATION CURVE FOR QUALITY CONTROL

In industrial QC usually one is not primarily interested in knowing the actual numbers of organisms but in determining whether specific samples fall within established specifications.

By drawing a horizontal line corresponding to the permissible microbial limits (10^4/g in the example of Fig. 5.5), the user can divide all points on the graph into those which fall above or below the established limit as determined by the conventional method (SPC). The intersection of this line with the regression line will identify the IDT most effectively dividing acceptable from unacceptable samples (permissible detection time). According to whether their detection times are shorter or longer than the permissible detection time, unknown samples can be quickly classified as acceptable or unacceptable.

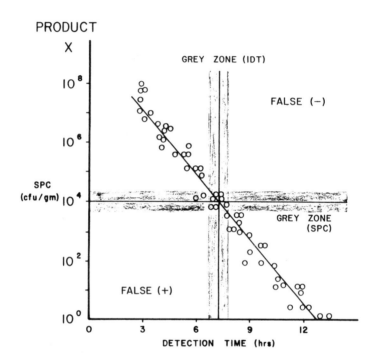

Fig. 5.5 A scattergram for the hypothetical product X, showing the false (-) and false (+) quadrants.

The samples falling in the upper left and lower right hand quadrants (Fig 5.5) are classified as acceptable or unacceptable by both techniques. The lower left and upper right hand quadrants represent disagreement between the impedance measurement and the SPC method. Although misclassified samples can theoretically occur at any point in the scattergram, an evaluation of their distribution around standard curves reveals that the majority represent samples having bioburdens close to the established limits, and detection times approximating the selected cut off times. Therefore, there is a time interval on either side of each curve where an uncertainty exists as to whether the level of organisms is above or below the permissible level. This time interval has been designated as the "grey zone" for the impedance method. Samples having marginal SPC (e.g., 8×10^3 or 1.5×10^4 in the example of Figure 5.5) are difficult to classify correctly since inaccuracies exist in the SPC method (Section 5.3). Therefore, a similar grey zone might be constructed for the plate count method.

The user can define the grey zone for each product and need retest only those samples whose IDT's fall within it. The width of a grey zone i.e., the time span on each side of the permissible detection time, in which retesting is indicated, is determined by both the tightness of points around the standard curve and the level of agreement required. For example, if we accept an error of one standard deviation (see appendix A for the method of calculation) from each side of the permissible detection time shown in Fig. 5.5 the grey zone will span from 6.7 to 7.2 hours. In this example all the samples that showed disagreement between the two methods will be included in this grey zone.

The slope (B) obtained from the calibration curve allows the generation time (tg) of the detected organisms to be calculated. It was shown in Chapter 3.5.2 that:

$$tg = \frac{\log 2}{B}$$

It is important to compare the generation time calculated in this manner with the value obtained previously for typical organisms from the product (see Chapter 4 for the method of calculation). If a large discrepancy is found between the two values, further investigation is required.

When the organisms have a very short generation time (e.g., <10 minutes) under the experimental conditions, the slope will be extremely steep (e.g., 1.8). With such a steep curve, a difference of just one hour in IDT will reflect a change of almost 2 logs in the standard plate count. Since observed standard deviations of IDT are generally in the range of 0.25 to 0.50 hour, it will be difficult to distinguish between samples differing by less than 0.5 to 1.0 log cycles in total count. For this reason conditions favoring exceptionally short generation times should be avoided.

The Y intercept value (A) is affected mainly by the conductivity of the medium and the duration of the lag phase (see Chapter 3). Since the range of media conductivity is limited and the lag phase usually occupies less than 1 hour, values for A are generally within the range of 6.5 to 8.5. An unusually high intercept value suggests that the lag phase is prolonged. The cause may be sub-lethal injury of the organisms or the presence of inhibitors in the system. An unusually low intercept value (i.e., <6.0), may indicate that those organisms detected in the impedance method are different from the predominant populations appearing on the plates.

5.8 FOOD APPLICATIONS

The calibration curves described above may be prepared for any product. Therefore, the impedance method could be used to estimate numbers of organisms in any product. However, proper development of the analytical protocol is the key to success. The importance of this will be demonstrated as we examine the protocols used to estimate impedimetrically total numbers of bacteria in raw milk. Applications to other foods are also discussed.

A. DAIRY

Several investigators (Gnan and Luedecke 1982; Cady et al. 1978b; and O'Connor 1979) have attempted to develop an impedimetric method for the estimation of total count (SPC) in raw milk. All workers mixed the milk (at a 1:1 ratio) with 10% yeast extract or BHI and incubated at 32°C. Inconsistent results and low correlation coefficients (below 0.8) were obtained in these studies. Gnan and Luedecke (1982) observed that samples with a high SPC and low IDT's were associated with entirely different flora than those with low SPC and short IDT's. They speculated that the impedance method detected psychrotrophs in the former case and mesophiles in the latter. In our recent research (Firstenberg-Eden and Tricarico 1983) it has been shown that mesophiles often have generation times four to five times shorter than that of psychrotrophs (Fig. 5.6). From the data shown in this figure it appears that 18°C is a better temperature for the impedance measurement since it allows similar generation times of psychrotrophs and mesophiles. It was shown that the growth on an agar surface resulted in shorter IDT's and better curves (Section 4.4). Therefore, 0.1 ml of raw milk was pipetted onto wells prefilled with modified plate count agar (MPCA), and the wells were incubated at 18°C. This procedure was tested with 110 milk samples from 14 different farms consisting of a varied ratio of mesophiles to psychrotrophs. The result (Fig. 5.7) shows a high correlation (0.96) between SPC at 32°C and IDT's at 18°C.

The usefulness of the grey zone can be demonstrated on this scattergram. If we choose a permissible level of 1×10^5 CFU/ml, the corresponding IDT will be 17 hours. By adding one standard deviation to each side of the IDT the time axis can be divided into three sectors: (i) 0 - 15.2 hours in which all the samples should have more than 10^5 CFU/ml; (ii) >19.1 hours where all the samples have less than 10^5 CFU/ml; (iii) 15.2 - 19.1 hours in which we have samples that cannot be classified with confidence. Using this scheme only one sample out of 110 was misclassified. This impedance procedure for total count was adopted by the Technical Committee preparing the 15th Edition of "Standard Methods for the Evaluation of Dairy Products" and will appear in this book as an alternative method to the SPC.

Since mesophiles are the predominant group of organisms present in farm fresh raw milk (Sogaard and Lund 1981; Zall et al. 1982), a method which estimates mesophilic load would be indicative of total count. Impedance monitoring at 35°C on the agar surface (MPCA)

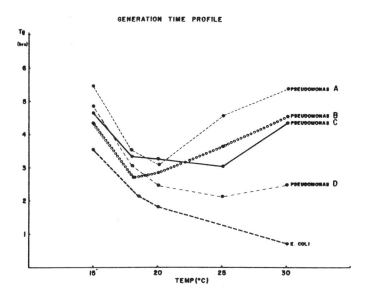

Fig. 5.6 Generation time profiles of four selected psychrotrophs and one selected mesophile over a temperature range 15-30°C (from Firstenberg-Eden and Tricarico 1983).

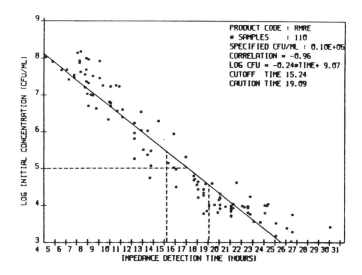

Fig. 5.7 Scattergram relating impedance detection times to total plate count for raw milk samples (from Firstenberg-Eden and Tricarico 1983).

was well correlated to the number of mesophilic organisms present in the sample. For all samples containing a majority of mesophiles, the IDT's at 35°C correlated well to SPC at 32°C. This procedure allows for a rapid screen of raw milk samples. Any contamination greater than 1×10^5 mesophiles/ml can be detected within 4 hours.

A collaborative study was carried out to establish the reproducibility of this impedance method in predicting counts of raw milk (Firstenberg-Eden 1984). Frozen and fresh raw milk samples were sent to six laboratories to be examined by standard plate count method (SPC) and by the impedance method which produced Bactometer Predicted Counts (BPC). The impedance results showed less variability (although not significantly so) among laboratories and between split samples in all three trials. In one trial the means of BPC and SPC were not significantly different. In another trial the mean BPC was 0.27 log units lower than the mean SPC. However, in this trial, the extreme differences between laboratories counting the same sample were around 0.42 log units. The results from this collaborative study indicated that the suggested impedimetric procedure yielded more reproducible results than the SPC. The results are obtained in a few hours compared with 48 hours for the SPC.

The psychrotrophic numbers in raw milk samples were determined by pipetting 0.1 ml of raw milk into a well containing 0.5 ml of solidified MPCA. The inoculated wells were preincubated at 10°C for 7 hours followed by impedance monitoring at 18°C (Firstenberg-Eden and Tricarico 1983). This impedance method correlated well (0.96) with psychrotrophic plate count (7°C for 10 days), and allowed detection of samples with more than 10^4 psychrotrophs/ml in less than 25 hours.

B. MEAT

A high correlation (0.97) between IDT's and CFU was also obtained for raw meat samples (Firstenberg-Eden 1983). In Chapter 4.8 we discussed in detail the effects of the preparation of the meat samples before the impedance test on quality of curves and detection times. The high correlation could be obtained only after the appropriate procedure for sample preparation was used.

C. FROZEN VEGETABLES

This work was done on the earlier version of the Bactometer (B32). Hardy et al. (1977) tested 257 samples of frozen vegetables. They found that the agreement between impedance and SPC in distinguishing samples containing more than 10^5 cell/g was 92.6% for 257 assorted frozen vegetables and somewhat higher (93 to 96%) when separate cut off times were used for each type of vegetable. Although their detection times related to the initial concentration there was a considerable spread in this data. They speculated that part of the spread was due to different growth rates and/or different durations of lag phase of the organism contained in these samples. It is quite likely that, if temperature and medium had been optimized to minimize the spread in generation times, better correlations could have been obtained with this type of product.

D. FISH

This work was carried out on the Malthus Microbial Analyser (Gibson and Ogden 1980). By using a temperature of 20°C (the temperature used normally for the SPC assay) and nutrient broth as the growth medium a correlation of 0.83 was obtained between the impedance results and standard plate counts. Most results of the assay were obtained within one day as compared with the 5 days required for the conventional method. The use of trimethylamine-oxide (TMO) medium (Wood and Baird 1943) resulted in a better correlation with SPC (0.89). It has recently been shown that fish spoilage bacteria can reduce TMO to trimethylamine (Easter et al. 1982). The TMO operates as a terminal electron acceptor under the microaerophilic growth conditions that exist in ampoules used in the Malthus apparatus.

5.9 CLINICAL APPLICATIONS

Estimation of bacterial levels in the clinical field was studied only in connection with urinary-tract infections. Since the companies producing instrumentation for impedance have concentrated their efforts in the food area, few clinical studies have been reported in the literature and none recently.

Lamb et al. (1976) used an electrochemical method developed by Wilkins et al. (1974), based on measuring a change in potential between two electrodes for the detection of urinary tract infections. This electrochemical detection method used reference and platinum electrodes placed directly into broth, detecting the presence of viable bacteria by measuring the change in voltage. This method differs somewhat from the impedance method described in this book. The results obtained in this study are shown in Fig. 5.8. Although positive urine specimens (counts $>10^5$) were detected

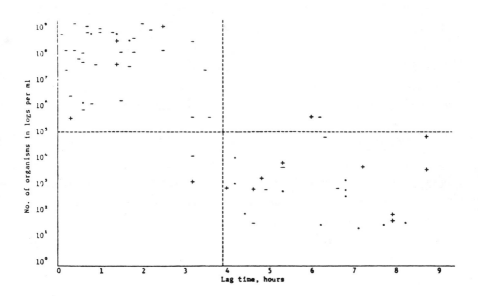

Fig. 5.8 Comparison of bacterial concentration in positive urine specimens and length of the lag phase. (+) = Gram positive; (-) = Gram negative; (·) Mixed bacteria (from Lamb et al. 1976).

rapidly using this method, there was quite a scatter in the results. When a cut off time of four hours was chosen there was 90% agreement between this electrochemical method and the standard method.

The early Bactometer (B32) was also used to detect urinary tract infections (Zafari and Martin 1977; Throm et al. 1977 and Cady et al. 1978). In the first two studies no scattergram is given for the relationship between IDT and log CFU. However, when a cut off time of 2-2.5 hours was chosen an agreement of 94-97% was obtained. Hadley and Senyk (1975) show a scattergram (Fig. 5.9) for the relationship between initial concentration of bacteria in 74 different urine specimens and IDT. Most of the specimens contained E. coli, but other organisms such as Proteus mirabilis, enterococci, Pseudomonas aeruginosa, and Enterobacter sp. are represented as points in this study. All of the cultures which initially contained 10^5 bacteria were detected within 2.5 hours. A more thorough study, with the same methodology, involving 1133 urine cultures was carried out by Cady et al. (1978). By defining an impedance positive culture as one that gives detectable impedance changes within 2.6 hours, 95.8% of the samples were correctly classified. In none of these studies was the medium or temperature optimized for the impedance method. It is conceivable that better results could have been obtained under optimized conditions.

5.10 SELECTIVE MEDIA

The detection of pathogenic microorganisms and their toxins in many products is impractical in many quality control laboratories. This is because the methods involved may be unreliable, especially when pathogens are sparsely or unevenly distributed in a product heavily contaminated with other organisms. Even when a sensitive method is available, it often requires expertise and equipment that is not available in such laboratories. Such difficulties have led to the widespread use of groups of organisms, which are more readily enumerated, as indicators of exposure to conditions that might introduce hazardous organisms and/or allow proliferation of

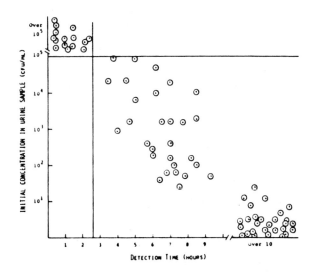

Fig. 5.9 Scattergram comparing the initial concentration of organisms in urine specimens with the time of detectable impedance change caused by the microorganisms. Samples (0.5 ml) of urine were inoculated into 0.5 ml of BHI (from Hadley and Senyk 1975).

infectious or toxigenic species. These groups are called indicator organisms. Examples of such groups are coliforms, enterococci, yeasts and molds.

The impedance method can be used, with the appropriate selective media, to estimate the numbers of such indicator organisms. However, it is important to keep in mind that the medium and temperature combination used in the conventional method might not be appropriate for the impedance method. Two examples of the use of selective media will be shown to demonstrate the importance of the appropriate medium-inhibitors-temperature combination for the successful use of impedance in estimating numbers of indicator organisms.

A. COLIFORMS

Fig. 4.2 (shown in the previous chapter) illustrates that the coliform medium (CM) especially developed for the impedance method resulted in better impedance curves than any conventional medium. The inhibitor level and the chosen temperature (35°C) minimized the generation times of the coliforms tested (Firstenberg-Eden and Klein 1983). In this study the coliforms had an average generation time (tg) of 26 min, 90% being in the range of 20-30 minutes. It is interesting that the slopes of the calibration curves for coliforms in a variety of food products were in the range of -0.61 to -0.76 (Firstenberg-Eden and Klein 1983; Firstenberg-Eden et al. 1984). This means that the generation time of detecting coliforms in the products was between 24 - 29 minutes.

The IDT's from CM medium correlated better with the counts of confirmed coliforms (in Brilliant Green) Violet Red Bile Agar (VRBA) than with the original VRBA data, suggesting that CM is more selective for coliforms than VRBA (Firstenberg-Eden and Klein 1983). The results of this study indicated that the impedance technique using CM medium was an effective method for determining coliform levels in beef samples, with significant time and labor saving over conventional methods.

The same medium was tested with a variety of dairy products (Firstenberg-Eden et al. 1984). All pasteurized dairy products tested in this study had very low coliform concentrations (<1 coliform/ml). Due to the low number of coliforms naturally present in these products, it was difficult to assess their numbers by either plate counts in VRBA or impedimetric detections in CM. Classifying these samples as meeting or exceeding the specified permissible level of coliforms seems to be the best approach.

Abusing the pasteurized products (or raw milk) increased the coliform counts, resulting in good correlations between IDT's in CM and counts in VRBA. Since similar regression lines were obtained for all the products tested, we could combine the results and conclude that an IDT shorter than 9 hours is indicative of coliform levels >10/ml while an IDT longer than 12 hours is indicative of levels <10/ml. Since there is an error associated with both methods, i.e., scatter of points around the regression line, a sample detecting between 9 and 12 hours cannot be classified with confidence (95% level) as having above or below 10 coliforms/ml. For raw milk, which usually contains a higher level of coliforms, an IDT of less than 6 hours indicates a coliform level $>10^3$/ml.

It has been found that confirmation of typical colonies from VRBA plates in BG is necessary when a product is first examined by these two methods. Raw and pasteurized milk necessitated confirmation and correction of the VRBA plate counts in order to obtain a satisfactory correlation between the two methods. Many colonies on VRBA initially classified as coliforms were shown in this way to actually be non-coliforms (Fig. 5.10). Confirmation was unnecessary in pasteurized heavy cream and ice cream mix for which a large percentage of samples produced a 100% confirmation (Fig. 5.11).

The pasteurized heavy cream required a modification to the growth medium to alleviate a drift problem in the impedance curves (Fig. 5.12). The addition of 0.1 M Tris buffer at pH 7.0 to CM improved the curves. The mechanism for this reduction in drift is not known although it has been shown that Tris increases membrane permeability (Irvin et al. 1981). The addition of Tris buffer to the medium may have altered the permeability of the fat globule membranes in the product, and this may have altered the impedimetric drift. Tris, an organic cation, also neutralizes negative charges in solution, (Schindler and Teuber 1978) which may have affected the drift of the impedance curve of cream in CM. Whatever the reason for improved curves, IDT's in CM + Tris buffer correlated well with levels of coliforms.

B. YEASTS

Conventional mycological media were developed to ensure colony formation (biomass production) over a period of 3 to 5 days. There was no reason to believe that these same media would necessarily be optimal for impedance measurements, which depend directly on the accumulation of charged end products of metabolism. It was,

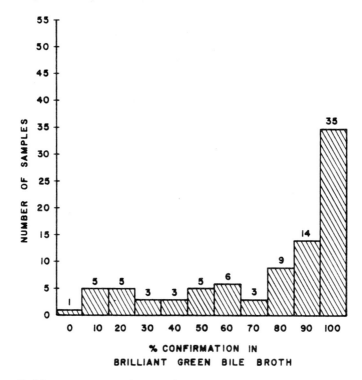

Fig. 5.10 Percent of confirmed coliforms from raw milk in Brilliant Green Bile broth.

therefore, not surprising that the medium chosen for impedance (CBAS) included a component (Yeast carbon base) that is usually used in taxonomy rather than enumeration, (Wickerham 1951). The second component of CBAS, ammonium sulfate, was included as a general nitrogen source for yeasts. Better impedance signals were obtained from autoclaved media than from filter sterilized media. Presumably, improved impedance responses in the autoclaved medium were the result of one or more heat-catalyzed reactions. Several workers have reported similar results, particularly in media with glucose and phosphate which are present in CBAS.

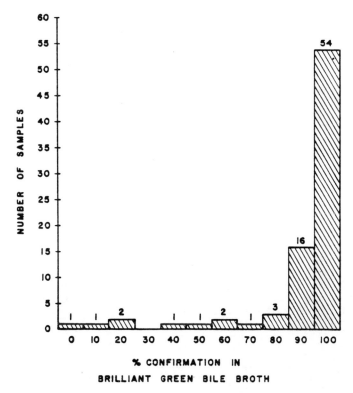

Fig. 5.11 Percent of confirmed coliforms from pasteurized heavy cream in Brilliant Green Bile broth.

The detection of yeasts in CBAS was best accomplished by measuring changes in capacitance, since capacitance signals provided stronger responses than the corresponding conductance signals, as shown in Fig. 5.13 (Zindulis 1984). This observation was consistent for all yeasts in all media tested. CBAS was superior to six common mycological media for the impedimetric (capacitance) detection of all the yeasts tested. An example of the media comparison using Kloeckera apiculata is shown in Figure 5.14. Yeasts in CBAS

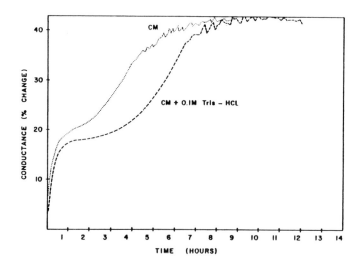

Fig. 5.12 Impedance curves from pasteurized heavy cream in CM showing the improvement in drift in CM + 0.1M Tris-HCl buffer (from Firstenberg-Eden et al. 1984).

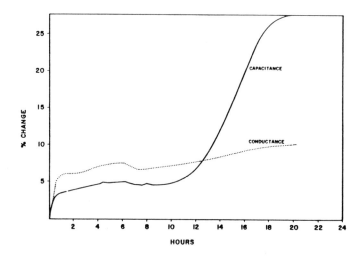

Fig. 5.13 Comparison of capacitance and conductance signal generated by Saccharomyces cerevisiae in CBAS (from Zindulis 1984).

produced flat or slightly negative baselines, followed by smooth, strong, positive acceleration. All other mycological media resulted in curves with weaker accelerations.

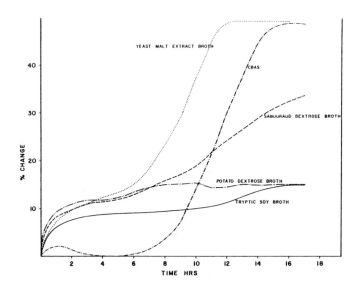

Fig. 5.14 Comparison of capacitance signals of Kloeckera apiculata in common mycological media and in the medium developed for impedance (CBAS) (from Zindulis 1984).

This medium was successfully applied for the detection of yeasts in orange juice (Zindulis 1984) and yogurt (Zindulis 1984B). A concentration of 10^2 yeast CFU/ml in a 1:10 dilution of orange juice in CBAS was detected within 25 hours by impedance, in contrast to three to five days by plate count. The use of a similar procedure to detect yeast in yogurts resulted in detection times shorter than 20 hours for 10^2 yeast CFU/ml.

Certain brands of yogurt or orange juice required a modification of CBAS. When a 1:10 dilution of the product in CBAS resulted in a pH below 4.5, addition of ammonium hydroxide to a pH of 5.0 - 5.5 was found to improve the impedance signal generated by yeasts. It was found that more favorable pH for yeast metabolism resulted in earlier detection times. The primary improvement in signal quality was reduced drift.

CHAPTER 6
Utilization of Impedance for Non-Counting Application

6.1 INTRODUCTION

In the previous chapter it was assumed that the impedance method is justified as long as it correlates with estimated CFU (Colony Forming Units) of microorganisms as determined by the traditional plate count method. This is indeed the situation, at least in food microbiology. However, it is important to understand that colony counts are not an end in themselves. Sharpe (1979) argues that plate count is a unique datum; nothing in the physical, chemical, biochemical or immunological world corresponds to it. It provides very little information about the ability of food to deteriorate, become toxic, or infectious. The reasons that plate counts were used to assess spoilage or potential hazard to health were because better methods were unavailable. The impedance method, which measures microbial activity and kinetics rather than numbers, can provide a better measure of acceptability and health hazard than the plate count method. Therefore, it is logical that impedance should be evaluated for its potential as a direct measure of product shelf-life. Although this seems to be one of the most promising uses of impedance, only a few studies have taken this approach to date. They include the shelf-life of pasteurized milk (Bossuyt and Waes 1983, Bishop et al. 1984) and spoilage of fish (Gibson and Ogden 1980).

One of the simplest applications for impedance in microbiology is the detection of organisms in a sample that is usually sterile.

Reported research on this application includes the monitoring of blood cultures (Throm et al. 1976, Specter et al. 1977, Throm et al. 1977, and Holland et al. 1980), spinal fluid (Kahan et al. 1976) and cosmetics (Kahn and Firstenberg-Eden 1984).

More complex applications involve procedures in which each sample occupies several channels in the instrument. The various channels simultaneously provide information about growth under different conditions. Impedance monitoring is easily multiplexed so this requirement can be readily met. Procedures requiring multiple channels include antibiotic susceptibility testing, determination of minimal inhibitory concentration (MIC), identification of organisms, phage typing and bioassays.

Further use for impedance is in the determination of microbial kinetics. Both microbial activity and bacterial generation times can be easily calculated from the impedance data (see Chapter 3). Only limited utilization of this type of application has been reported to date.

6.2 SHELF-LIFE

An important concern of manufacturers of perishable food is to obtain a rapid method to determine the potential shelf-life of their products. In the dairy industry several changes, particularly the trends toward wider marketing areas and less frequent deliveries, have greatly increased the importance of the potential shelf-life of the product. Initial bacterial counts have proven to be of limited value in predicting the keeping quality of milk and other food products (Hankin and Anderson 1969, Hankin and Stephens 1972, Hankin et al. 1977, Randolph et al. 1965, Watrous et al. 1971). The reasons for this are illustrated by Table 6.1. Milk that contains 10 organisms per quart immediately after packaging can have as low as 350 organisms/quart after five days at 7°C. However, in this same period the same initial count can result in 10^7 organisms/quart and a spoiled product if the bacteria multiply faster. Therefore, shelf-life is likely to correlate better with bacterial activity than with its initial concentration. Since the impedance method measures metabolic activity and, as a matter of course integrates the effects of numbers of organisms and their metabolic activity, it would seem to be a most suitable tool for shelf-life prediction. It is therefore surprising that only a few studies are reported in the literature on the use of impedance for this purpose. Most of these reports are about prediction of shelf-life of pasteurized milk.

Bossuyt and Waes (1983) showed impedance measurements to be useful for the detection of post-pasteurization contamination of milk by gram-negative bacteria. Martins et al. (1982) concluded that for prediction of the shelf-life of milk on the day of pasteurization, the impedance method is superior to both the standard plate count and the psychrotrophic bacterial count. They did not determine, however, how accurately the impedance method predicted the shelf-life of pasteurized milk.

TABLE 6.1 The Effect of Generation Time and Storage Time on Number of Bacteria Presented in a Quart of Milk Due to Ten Contaminants per Quart (data adopted from Blankenagel 1976)

Generation Time (Hr)	Number of Generation Times in		Final No. of Bacteria/Quart	
	5 Days	7 Days	5 Days	7 Days
8	15	21	350	22,000
6	20	28	11,000	290,000
4	30	42	11,000,000	47,000,000,000

In a more recent study Bishop et al. (1984) investigated potential predictors of shelf-life for 100 pasteurized whole milk samples obtained from retail outlets and from dairy processors. Parameters studied were: organoleptic evaluation, standard plate count, psychrotrophic bacterial count, modified psychrotrophic bacterial count, Moseley test, and impedance. Correlation coefficients were obtained for all possible data combinations in an attempt to detect significant relationships between the parameters studied and the actual shelf-life of the product (Table 6.2). None of the direct counts correlated well enough with shelf-life to allow satisfactory prediction. The Moseley test showed an acceptable relationship to shelf-life, with a correlation coefficient of 0.84. However, 7 days were required to obtain results. The impedance method proved to have the most significant relationships to shelf-life with a correlation coefficient of 0.88.

Table 6.2 Correlation Between Shelf-Life and Microbiological Tests

Method	Correlation to Shelf-Life	
	Linear	Quadratic
Impedance (21°C, up to 40 hr)	0.87	0.88
Moseley Test (5 Day + 48 hr at 32°C)	−0.84	−0.84
Plate Count (32°C, 48 hr)	−0.52	−0.56
Psychrotrophic Count (7°C, 10 D)	−0.58	−0.64
Modified Psychrotrophic Count (18°C, 18 hr followed by 21°C, 25 hr)	−0.63	−0.67

Since any test for keeping quality of milk must emphasize those organisms (usually present in low numbers) that can grow quickly and spoil the product in the refrigerator, a preincubation step was included in the impedance method. Preincubation of diluted milk (diluted 1:1 in plate count broth) for 18 hours at 18°C resulted in quick growth of spoilage organisms. This preincubation step was followed by the addition of 0.5 ml of the diluted milk onto the surface of MPCA and monitoring impedance at 21°C.

Using this method the results are available in 25-38 hours (18 hour preliminary incubation + 7-20 hour impedance detection). The impedance method provides a result 6-8 days before the Moseley test. In addition the procedure requires much less work than the Moseley test.

6.3 STERILITY TESTING

The aim of this test is to verify that there are no viable microorganisms in a given volume of product. For this purpose it is less important to equalize the microbial generation time. The first step is to develop a medium (or a number of media) that will assure the detection of all organisms that might appear in such product. In the next step the maximum time necessary to monitor impedance is determined. This growth period should be long enough to allow the lowest levels of the slowest growing organisms to be detected. In this case, any detection indicates contamination. In most sterility tests low levels of organisms have to be detected. A concentration step is very important to obtain shorter IDT. This type of application was most widely tested in blood cultures.

A. BLOOD CULTURES

Rapid detection of bacteremia is a major goal in clinical microbiology laboratories. The high index of suspicion for bacteremia in critically ill, febrile patients, dictates that the number of cultures drawn will far exceed the number of cases of bacteremia uncovered. Thus, the clinical laboratory must process ten cultures for every one found positive, frequently employing two, three, and even four culture bottles per specimen. Obviously such a large number of bottles can only be examined periodically. A positive culture which gives visual indications of growth (turbidity or color change) shortly after being examined will go undetected until the next scheduled observation and thus frequently a day is lost. Hourly inspection would certainly speed results, but would be impractical without some form of automation. Since impedance monitoring is automated and employs continuous measurement, it is a prime candidate for the job.

Generally, impedance monitoring yields results much more rapidly than conventional methods. This is not due so much to the sensitivity of impedance measurements (its thresholds are comparable to optical systems). In part it is due to its freedom from optical detection and the extra incubation needed before microbial growth can be detected visually against the opaque mixture of blood and medium. In part it also results from the capability of impedance measurements to monitor continuously, rather than intermittent monitoring. This allows the operator to be warned the instant a reliable change is detected.

Very low inocula of some slow-growing organisms, which took many days to show visual changes, took just as long in the impedance system (Cady 1976). In those cases, impedance monitoring at least offered considerable savings in the labor usually required to make the numerous observations that such slow-growing cultures require. Of greater value, perhaps, are the most rapidly growing cultures (which contain the largest initial inocula and, therefore, are of the greatest clinical concern), which are also most rapidly detected by the impedance system. Conversely, the cultures containing only a few slow-growing organisms, and in which the clinical significance may be controversial, are detected least rapidly (Cady 1976).

All of the bacteria examined by Specter et al. (1977), including __E. coli__, __Serratia__, __Klebsiella__, __Pseudomonas__, __Staphylococci__ and __Streptococci__ gave relatively rapid changes in impedance; growth was usually detected within 6-18 hours. In a limited clinical trial, the Bactometer (B32) detected the one positive specimen from 40 clinical blood specimens as rapidly as by radiometric measurement. The impedance measurements were found to be as accurate as those from the radiometric method and detection times were similar (Specter et al. 1977, Cady 1976). Most false negative cultures (organisms detected upon subculture but not detected impedimetrically) can be avoided by the use of several culture bottles containing different media (Cady 1976).

Two interesting studies were conducted with non-commercial systems. Holland et al. (1980) used a prototype system constructed by Abbott Diagnostics, which measures changes in voltage. Overall microbial growth was detected an average of 18 hours earlier by their instrument than by the conventional method; however ca. 10% inoculated cultures showed visible microbial growth before characteristic electronic response was detected. In a limited evaluation of 156 clinical specimens a complete agreement was obtained between the electronic method and standard method.

Kagan et al. (1977) constructed an impedance chamber and used it in conjunction with the lysis-filtration technique (Zierdt et al. 1976). With this technique the organisms in 14 ml of blood were concentrated on a filter and the antibiotics were washed away. This in turn resulted in earlier detections and increased overall sensitivity. The lysis-filtration impedance detection technique detected 49 (92%) of the positive cultures. Two of the false negatives were from a patient with Haemophilus influenzae bacteremia. The Brain Heart Infusion, used for the impedance, did not support the growth of H. influenzae. The conventional broth method detected only 30 (56%) of the total number of positive cultures. In one of the cases investigated the patient had prolonged bacteremia as a result of a Corynebacterium infection which was extremely resistant to drugs. This organism grew quite slowly, with very little visible reaction in the culture bottle. Impedance detection took 24 hours, whereas the bottle in the conventional technique showed no growth until 72 hours. This bottle remained clear, though a blind 24 hour routine subculture showed growth at 72 hours. Organisms grew more "strongly" in the lysis-filtration impedance technique than in the standard technique especially after antibiotic therapy was begun.

B. "INDUSTRIAL STERILITY"

In some cosmetic, pharmaceutical and food products a "sterility test" is performed. Usually in these products there is no attempt to totally eliminate microorganisms, at great cost to the customer, but rather to ensure that microorganisms are at low enough levels to prevent risk to the customer or spoilage of product. Occasionally the product becomes contaminated, often with high numbers of organisms. The impedance method can provide early warning of such contamination by performing what might be called an "Industrial Sterility" test. The idea is that, since there is a large difference in microbial concentration between "good" and "bad" samples, and we are usually interested in a "yes" or "no" answer, we can test for the presence of organisms in a certain portion of the product. The volume of product chosen will be such, that the normal product would contain less than one organism. Therefore, only if the product is contaminated above the permissible specified level will detection occur. For example if a mascara normally carries between 1 - 20 organisms/g and its specifications call for no more than 1,000, a 1:50 dilution of the product will assure the detection of samples which are out of spec, while resulting in no detection for normal samples. The keys for the success of such procedures are:

(i) A large difference in contamination level must exist between "good" and "bad" products.

(ii) An incubation time and medium must be found that will allow the growth of the flora to a detectable level from above spec samples.

This approach will be illustrated with data from several cosmetics and toiletry products. A variety of commercial creams, lotions and shampoos were tested. All the products had counts much lower than their specified permissible levels (500 - 1,000 organisms/g).

Therefore, all the products were inoculated with contaminating bacteria typical of cosmetics. The product was diluted in Eugonic broth containing neutralizers (Tween 80, Triton X-100, and lecithin). A portion of the broth was added to a vessel containing modified plate count agar (MPCA), and impedance was monitored. None of the normal samples gave positive plate counts, nor did they cause an impedance change. The products were tested with twelve bacteria, including seven Pseudomonads. "Sterility" could usually be assured within 24 hours. The chosen medium (MPCA) aided the growth of Pseudomonas and eliminated interference from the product (Kahn and Firstenberg-Eden 1984).

A floor cleaner (pH 11.0), naturally contaminated with a Pseudomonas, required a different procedure to test sterility. The Pseudomonas grew only at an elevated pH; changing the pH of the medium optimized its growth and impedance response. Brain Heart Infusion broth at pH 9.0 was used as both diluent and growth medium for this test. The impedance method than detected this contaminant in the floor cleaner (1×10^4 CFU/ml) in 17.6 hours. This again demonstrates the importance of an appropriate medium for the impedance test.

6.4 SENSITIVITY TO ANTIBIOTICS

Measurement of the in vitro response to antibiotics of various pathogenic microorganisms is an attempt to predict rapidly the in vivo response, which is not amenable to controlled experimentation for obvious reasons. The ideal information one would wish to obtain from in vitro studies includes the responses of the organism across the possible range of its in vivo concentrations, and across the range of antibiotic concentrations achievable with the available therapeutic regimens. Further, the ideal data would include the effects of duration of exposure of pathogen to antimicrobial agent. This is a lot to ask of a test procedure and in practice only a portion of this information is obtained.

Traditional antibiotic susceptibility test methods usually involve a number of technical steps followed by overnight (Ca. 18 hour) incubation, after which a visual assessment of the end point is made. A case can be made for continuous monitoring of antibiotic response. For example, suppose one exposes organisms to a low concentration of penicillin, such as one-tenth of the minimum inhibitory concentration (by definition, not inhibitory at the end of eighteen hours in the standard test). From the growth kinetics during the first few hours of exposure, it is frequently seen that the number of viable organisms declines rapidly to a minimum, before the population recovers and starts multiplying again. Explanations are readily available. Time may be required for the organisms to induce the appropriate enzymes required to survive (e.g. penicillinase), or to acquire the appropriate R factor. Perhaps only a portion of the population is genetically endowed with the proper enzymes. In any event, conclusions as to whether the organism is susceptible or resistant will depend on the incubation time allowed before the assessment is made. Similarly, high concentrations of organisms frequently appear to have higher minimum inhibitory concentrations than lower concentrations of the same organism. Therefore, there is a clear advantage to be gained from continuously monitoring growth, as with the impedance method, over any technique measuring only initial and final indices of growth.

When pure cultures of organisms were exposed to various levels of antibiotics and impedance was monitored continuously, different impedance patterns were observed. The effects of antibiotics on impedance curves could be divided into four groups:

1. <u>Shift in IDT</u>

   The effect of novobiocin on the <u>Salmonella</u> <u>typhi</u> patterns illustrates this phenomenon (Fig. 6.1). Increasing concentrations of novobiocin increased the IDT. For example, the IDT of the control was Ca. 4 hours, and 20 mg/l delayed it by 5.5 hours. However, the slopes after acceleration and the maximal change in impedance were similar for all these concentrations of antibiotics.

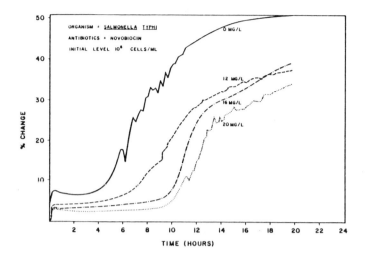

Fig. 6.1 Shift in detection times due to antibiotics.

2. Changes in Maximum

This type of response was obtained when 10^7 E. coli cells were inoculated into one ml of media with varying concentrations of ampicillin. The onset of impedance change did not vary, but the amplitude of the response diminished with increased concentration of the antibiotic (Fig. 6.2). Growth in the 1 mg/l sample peaked at about 6 hours and at a lower value than the control. Higher concentrations of antibiotics resulted in earlier and lower maxima. In this experiment the inoculum level was above the threshold level for impedance detection. Therefore, even though cell division had probably stopped, there was probably considerable metabolic activity. The period for which this activity continued and its intensity, was related to the concentration of antibiotic. Lowering the inoculum to 10^6 cells/ml required growth before detection occurred. At this inoculum level E. coli exposed to higher levels (>4 mg/l) of ampicillin were not detected. The maximal values attained with

2 mg/l of ampicillin were lower than the control, indicating growth to a lower concentration than the control.

3. Decrease of Slope after Acceleration

The slopes of the impedance curves, due to growth in the presence of the antibiotics, are shallow compared with the control. Increasing the antibiotic levels results in a decrease in the slope, until no acceleration is observed. This phenomenon is usually not seen by itself, but in combination with (1).

4. Combination of all Phenomena (1 - 3)

Fig. 6.3 shows the change in impedance patterns resulting from the addition of tetracycline to an E. coli culture. The lowest concentration of tetracycline (1 mg/l) simply delayed the onset of acceleration therefore increasing the IDT. In other respects the growth curve resembled the control without any antibiotic. At 2 mg/l tetracycline there was a further delay in

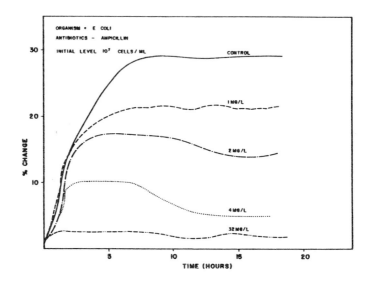

Fig. 6.2 The effect of antibiotics on the maxima attained by the impedance signal.

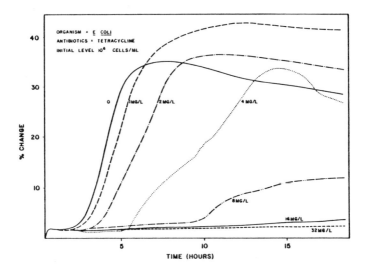

Fig. 6.3 Effect of antibiotics on shift in detection times and the reduction of slopes and maxima.

acceleration, but in addition, there was a pronounced decrease in the rate of response. This was increasingly noticeable at the 4 mg/l and 8 mg/l concentrations. At 8 mg/l the maximum was also reduced. Finally, 16 mg/l produced only a very gradual impedance increase up to eighteen hours, which may have been simply drift.

One could speculate that, if the organisms need only an adjustment period prior to multiplying normally in the presence of antibiotics, only effect (1) will be seen. If however, antibiotics affect the generation time of organisms, a combination of (1) and (3) will be seen. Therefore, impedance curves can provide much more information on microorganism-antibiotic interaction than has been hitherto available. It will be therefore of great interest to explore their applications in clinical chemotherapy.

The impedance technique showed disturbances in microbial growth at lower antibiotic concentrations than conventional techniques. This feature could make the technique useful in food microbiology for detection of low levels of antibiotics in products such as milk and meat.

To date very little has been done to explore the use of impedance measurements in either automated antibiotic susceptibility tests or determination of minimum inhibitory concentrations (MIC). Colvin and Sherris (1977) compared the effects of ten antibiotics on five clinical isolates by the traditional broth dilution test and impedance detection. Visual readings of minimal inhibitory concentrations (MIC) from overnight incubation were compared with those determined from IDT using inocula of 10^6 organisms/ml. Ninety-three percent of the results were within a single two-fold dilution of each other. However, this correlation was lowered to 34% when the cut off point of the impedance test was set at 6 hours. By increasing the initial electrical impedance inocula 10-fold, the correlation between the 6 hours impedance MIC and the overnight visual MIC was improved to 74%. Better correlations might have been obtained if curve patterns were taken into account.

Porter et al. (1983) reported that in 462 tests, there were only 13 discrepancies between the results of the disk diffusion tests (overnight) and 6.3 hour conductance measurements, using the Malthus system. In eight of these discrepancies the organism was reported as resistant to disk testing but sensitive to the conductance assay. When the disk diffusion test was repeated with lower inoculum density the organisms were considered sensitive. Porter et al. (1983) argued that if one were to use the antibiotics at the clinically attainable concentration and with an inoculum density commensurate with that found, for example, in urinary tract infections, the results of the conductance assay would be more relevant to the clinical management of the patient. These results would doubtless show a poorer correlation with those obtained with the disk diffusion test.

6.5 CHARACTERIZATION AND IDENTIFICATION

By measuring impedance changes associated with bacterial growth simultaneously in a number of different media, characteristic patterns can be obtained. Although little work has been done to date toward applying impedance measurements to bacterial identification, this field appears to have some potential because of the ease with which large numbers of channels (with different growth conditions) can be sampled. Cady (1978) showed (Fig. 6.4) curves demonstrating the feasibility of determining an organism's sugar assimilation patterns from impedance measurements. This figure shows the impedance change resulting from Neisseria gonorrhoeae inoculated into Thayer Martin broth, and Thayer Martin broth with maltose and sucrose substituted for the glucose. The fact that N. gonorrhoeae assimilated glucose, but no maltose or sucrose, was apparent within 10 hours.

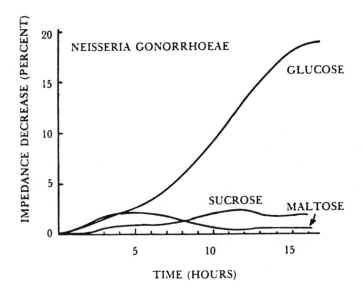

Fig. 6.4 Impedance changes resulting from Neisseria gonorrhoeae (10^7/ml inoculum) in Thayer Martin Broth and Thayer Martin Broth with its glucose replaced by sucrose and maltose (from Cady 1978).

An extension of this concept is to use growth inhibitors to build a repertoire of different media so that the results of growth in such a set of media can be used to further characterize and even identify an unknown organism. Firstenberg-Eden et al. 1983 have tried this approach to differentiate between salmonellae and non-salmonellae.

Analysis of foods for Salmonella is a common practice in the food industry. The conventional methods require many days, are labor intensive and require particular expertise in the identification scheme. Therefore, more rapid, less labor intensive, and automated methods are desired. Thirty eight Salmonella serotypes, all isolated from foods, were used in this study. Interfering flora were obtained by adding a variety of foods (dry milk powder, potato egg salad, eggs, poultry parts, etc.) to the pre-enriched medium, and transferring 10 ml of this solution into commonly used enrichment media. After the enrichment the predominant organisms were isolated on Salmonella-Shigella agar or Tryptic Soy Agar.

Although identification of Salmonella might require the determination of a variety of characteristics, a rapid automated method should preferably be based only upon a minimum number of tests. For the sake of economy it seemed feasible to use 4-5 tests for the presumptive identification of Salmonella. From typical biochemical reactions of Enterobacteriaceae (Edwards and Ewing 1972) the four most promising tests for the differentiation of Salmonella appeared to be: (i) fermentation of dulcitol; (ii) no fermentation of lactose; (iii) H_2S production; and (iv) lysine decarboxylation. An additional test, sensitivity to Felix 0-1 bacteriophage was included. Pure broth cultures of Salmonella and interfering organisms were diluted in peptone water. Approximately 10^4 cells/ml were added to sterile wells containing 1.5 ml of the appropriate media. The following results were obtained:

1. Dulcitol

Stewart et al. (1980) suggested the use of dulcitol fermentation as a presumptive radiometric test for Salmonella.

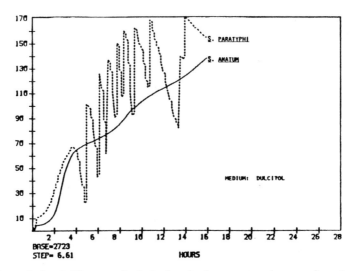

Fig. 6.5 Effect of dulcitol fermentation on impedance patterns.

Fig. 6.5 shows the capacitance curves resulting from the growth of two organisms, one fermenting dulcitol, and the other incapable of dulcitol fermentation. Gas production accompanying fermentation produced a visible disruption of the signal (see Chapter 4.4). Eighty one percent of the Salmonella serotypes fermented dulcitol. This result is similar to that of Edwards and Ewing (1972) but lower than the Australian data cited by Stewart et al. (1980). All the Salmonella that did not show gas on the impedance curves also failed to produce gas in fermentation tubes within 24 hours.

2. Lactose

The reaction tested was growth in a lactose medium containing bile salts and sodium desoxycholate as inhibitors, without evidence of gas production. Typical results of this test are depicted in Fig. 6.6. None of the Salmonella tested fermented lactose while most of the interfering organisms did (Table 6.2).

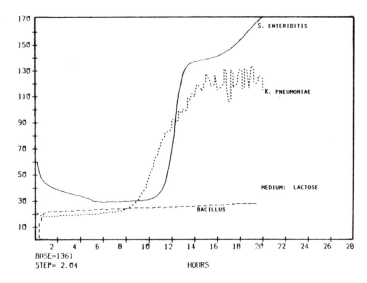

Fig. 6.6 Effect of growth and lactose fermentation on impedance patterns.

3. Lysine-decarboxylase

The test medium was prepared at a low pH (4.5 - 5.0). Organisms unable to decarboxylate lysine would be unable to grow to detectable numbers. Only three <u>Salmonella</u> did not decarboxylate lysine. However, many of the interfering organisms did grow in the medium. These organisms were also able to decarboxylate lysine in conventional media.

4. Hydrogen Sulfide Production

The impedance instrument was unable to detect H_2S production. By adding sodium thiosulfate and ferric ammonium citrate to one of the previous media (e.g. dulcitol or lactose), the blackening of module wells due to H_2S production could be seen after the removal of the module wells from the instrument. Since this reaction did not appear to contribute toward the classification of the organisms it was eventually dropped.

5. Phage Sensitivity

This last reaction makes use of the Salmonella-specific bacteriophage Felix-01 (Kallings 1967, Welkos et al. 1974). Fig 6.7 shows the impedance curves resulting from the growth of S. St. paul, with and without phage. The phage delayed the acceleration of the impedance signal. Such a delay was not observed with non-salmonellae (Fig. 6.8). The sensitivity to phage could be tested in a variety of media. For example, the addition of phage to one well of the dulcitol medium was used to assess both dulcitol fermentation and phage sensitivity (Fig. 6.9). Three Salmonella were not sensitive enough to the phage to cause a consistent delay in IDT. These Salmonella did not show plaques on plates. None of the interfering organisms were sensitive to the phage.

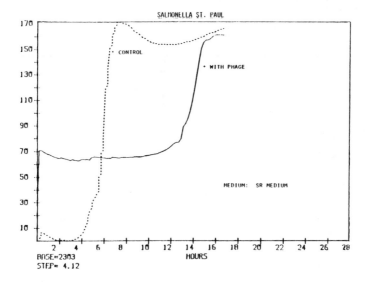

Fig. 6.7 The effect of sensitivity to Felix 0-1 bacteriophage on impedance patterns.

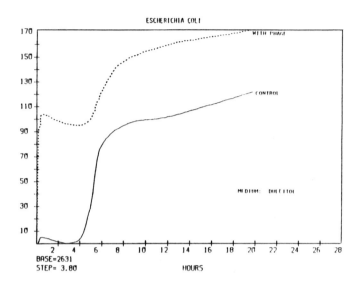

Fig. 6.8 The effect of Felix O-1 bacteriophage on non sensitive organisms.

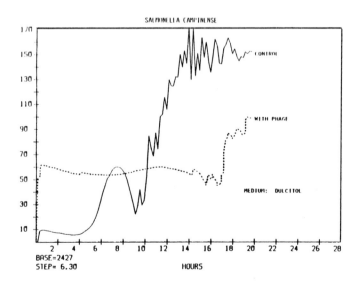

Fig. 6.9 Combined effect of dulcitol fermentation and sensitivity to Felix O-1 bacteriophage.

A summary of the probabilities of reaction for the 38 Salmonella isolates and the interfering organisms is shown in Table 6.3. A classifier was then developed that will distinguish between Salmonella and most other common food isolates. The classifier used is shown in this table. A positive value for g(x) indicates that the unknown organism is most probably a Salmonella. Values of g(x) between 0 and 0.3 identify the organism as suspect for being a Salmonella, while values less than -0.5 indicate that the organism in question is not Salmonella. All salmonellae tested (except S. gallinarium which had atypical biochemical characteristics) resulted in g(x) values of greater than -0.3. S. paratyphi had a g(x) value of -0.2, while the other 36 Salmonella serotypes had positive values. None of the interfering organisms had a g(x) value greater than -0.6.

Table 6.3 Probabilities of Positive Reactions

	Decarboxylate Lysine (i_1)	No Lactose Fermentation (i_2)	Dulcitol Fermentation (i_3)	Phage Sensitivity (i_4)
Pi (Salmonella)	91.4	99.0	81.1	91.9
qi (Interfering organisms)	43.3	16.7	16.7	1.0

$$g(X) = \sum_{i=1}^{4} [X_i \log \frac{Pi}{qi} + (1-X_i) \log \frac{100-Pi}{100-qi}]$$

$x_i = 1$ if positive

$x_i = 0$ if negative

6.6 Microbial Growth Kinetics

In foods, various determinants of microbial growth such as temperature, water activity a_w, pH, and concentrations of inhibitors affect the stability of the product. It would be advantageous to investigate the interactions between these various factors as they affect microbial kinetics, as we attempt to extend the shelf-life and safety of foods. In the development of starter cultures for the dairy, wine, or meat industry, for example, it is very important to determine in advance how active these cultures are going to be. These are only a few examples of the industry's need to assess growth rates and microbial kinetics. The impedance method provides a unique opportunity to study microbial kinetics and activity. Unfortunately, very little study has been made in this area to date.

Calculation of generation times by the conventional methods is time consuming. The optical density method requires lengthy calibration and does not result in very accurate data. It usually involves several measurements at pre-determined intervals. If the food suspension is not clear the optical method cannot be used, and one is left with the plate count method. In order to determine generation times one has to pour numerous plates at predetermined intervals. Both these methods result in less precise data than continuous methods.

Generation times can be easily calculated from impedance data. Several methods can be used to calculate generation times. In Chapter 3.5.2 it was shown that the slope of the calibration curve relating IDT to plate count can be used for this purpose. Another method was described in Chapter 4.5, where generation times were calculated from the difference in detection times between two different dilutions.

Microbial metabolic activity and generation times are not always simply related. Some starter cultures may multiply rapidly but produce little of the desirable acid, while another culture which multiplies slower may produce more. Calculation of microbial activity by standard methods is usually tricky. The metabolic activity of cultures can be calculated from impedance data as described in Chapter 3.5.3. Curve parameters such as slopes and maxima can also be used to assess microbial activity. For example, the metabolic activity of a dairy starter culture can be inferred from the steepness of the slope after detection. The greater the activity of the culture, the steeper is the slope. A standard curve can be produced relating percent change in impedance/hrs. to acid production.

The potential use of impedance to obtain a better understanding of microbial kinetics in complex systems is one of the most exciting areas for this technology. There is every possibility that it will provide new insight and understanding in many microbial systems.

CHAPTER 7
Instrumentation

7.1 INTRODUCTION

The need for improved instrumentation for analyzing microbiological samples combining speed and accuracy at a reasonable cost, is evident. This chapter is therefore dedicated to instrumentation, constructable by the laboratory, or available commercially, capable of implementing the concept of impedance microbiology as a practical and reliable tool.

As mentioned before, impedance is the resistance to the flow of an alternating electrical current. Cultures of microorganisms bring about changes in the chemical composition of the growth medium through the enzymatic activity associated with multiplication and metabolism. These chemical changes alter the impedance of the medium and provide an indirect measure of microbial growth and metabolism.

One of the advantages of impedance measurement is the ability to multiplex a single measuring circuit, so that a very large number of samples can be measured simultaneously without physical movement of sample, electrodes, or measuring system. As computer technology continues to develop, more powerful systems capable of handling more samples, and providing sophisticated data analysis are becoming increasingly feasible.

In principle, every impedance-based microbiological growth analyzer has at least three major <u>functional</u> units (Fig. 7.1): (1) electrobacteriological interface, (2) controller/analyzer and,

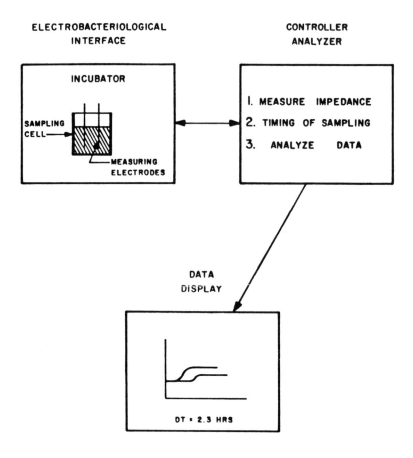

Fig. 7.1 The major components of an impedance based analyzer.

(3) data display. The first component includes the incubation unit which can be set to the proper temperature, and the electrochemical wells, i.e. containers (bottles or multi-well modules) with implemented electrodes. The function of this unit is to transform the metabolic activity of microbial cells into a measurable electrical quantity.

The second unit carries out the actual measurement of impedance and various analyses, such as the determination of detection times. In multi-well systems this unit also provides the control and timing for the measurements on each sample. The interface to the operator is accomplished through the third unit, in which either impedance curves and/or processed quantities are displayed and recorded.

In its simplest form a system may be constructed out of a single incubated sampling well, connected to a measuring circuit which displays the electrical variations as a function of time, for example, on a strip chart recorder. The next section discusses technical aspects of analog measuring circuits and is directed towards users interested in impedance microbiology, but who do not wish to purchase a commercial system. With the appropriate engineering support, a small unit can be easily constructed.

Section 7.3 deals with various aspects of a commercially available system, the Bactometer® Microbial Monitoring System M123 produced by Bactomatic, A Division of MTC, Princeton, NJ. A second commercial system, the Microbiological Growth Analyzer, produced by Malthus, is described in Section 7.4. The reader should note that a reputable manufacturer does not just sell an electronic gadget but a whole system in which the microbiological applications and protocols are inseparable and both aspects should be fully developed and supported.

7.2 IMPEDANCE MEASURING CIRCUITS

Bridge measurements provide the most direct way to compare unknown impedances with known standard impedances. The impedance of the sample cell may be represented by series (or parallel) combination of conductive and capacitive elements. The AC bridge illustrated in Fig. 7.2 consists of an alternating (sinusoidal) voltage source, two fixed resistors (R_1), a variable resistor R_B, a variable capacitor C_B and the sample cell represented by the series combination R_S and C_S whose values should be determined. At balance, i.e. $R_B = R_S$ and $C_B = C_S$, the voltage as well as the phase angle across the detector are nulled. Therefore, manual adjustments of both R_B and C_B are required until the null conditions are achieved. The unknown sample cell parameters R_S and C_S are then identical to the adjusted values R_B and C_B respectively. The series R-C combination of the sample cell can be also matched by a parallel combination $R_B' - C_B'$ illustrated in Fig. 7.3. To balance the bridge the variable resistor R_B' and capacitor C_B' will be set to the following values:

$$R_B' = R_S \left[1 + \frac{1}{(wR_S C_S)^2}\right]$$

$$C_B' = \frac{C_S}{1 + (wR_S C_S)^2}$$

where w is the angular frequency = $2\pi f$ and f is the source frequency in Hz. It might seem most convenient to employ the bridge with series R-C balancing, from which no further calculation is required. However, the parallel combination is used more frequently. For the range obtained with microbiological media, the parallel combination uses capacitors which are cheaper, more readily available, and more accurate than those required for the series combination.

Although bridges can provide very accurate impedance measurements, they are seldom used as actual instruments in impedance microbiology. The time required for the balancing procedure is impractical whenever large numbers of samples are monitored. In most applications very high accuracy measurements are not essential, and less accurate but fully automated circuits may be employed.

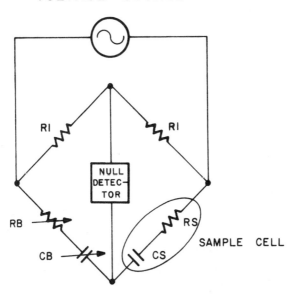

Fig. 7.2 A.C. bridge with the impedance of the sample cell balanced by a series R and C combination.

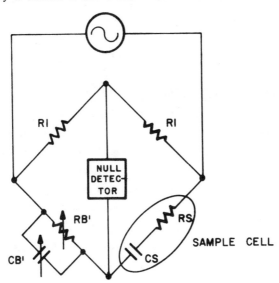

Fig. 7.3 A.C. bridge with the impedance of the sample cell balanced by a parallel R and C combination.

Simple electronic circuits can be used to measure either conductance or capacitance values. The basic configuration is illustrated in Fig. 7.4 where a single current source drives the sampling cell, represented by Zc. The same sinusoidal source is clipped by the shaper and multiplied by the voltage of the sampling cell. If the current source is $i \cdot \cos(wt)$, where $w = 2\pi f$ and f is the frequency in Hz, then the resultant cell voltage is $v \cdot \cos(wt + \phi)$ where $\cos\phi = R_c$ and $\sin\phi = \frac{1}{wc}$. If, now, the clipped voltage v_2 is in phase with the source, then the resultant multiplied wave v_1 is a combination of higher harmonic components (which are attenuated by the low pass filter) and a DC component which is proportional to $\cos\phi$, i.e. to the resistive part of the sampling cell R_c.

The same circuit can be applied to measure the capacitive part of the sampling cell C_c. In this case the square wave should be first shifted by 90° relative to the current source. The resultant DC component is now related to $\sin\phi$, which is inversely proportional to the capacitive part of the combination. The reader can also verify that the total impedance of the sampling cell may be measured by the same circuit by simply applying the cell voltage to the second input of the multiplier (v_2). Therefore, a relatively simple circuit may be utilized to obtain continuous readings of either impedance, conductance or capacitance values.

7.3 THE BACTOMETER® M123 MICROBIAL MONITORING SYSTEM

This system (Fig. 7.5), produced by Bactomatic, A Division of MTC, Princeton, New Jersey, is a multi-incubator system capable of accommodating up to four separate Bactometer processing units (BPU's). Each BPU contains two separate incubator compartments, each of which is capable of monitoring up to 64 separate samples. Therefore, a total of 512 samples can be monitored simultaneously, and up to 8 temperatures may be separately set. For test purposes, samples are pipetted into individual wells of Bactometer disposable modules (Fig. 7.6) or transferred to larger vials containing paired

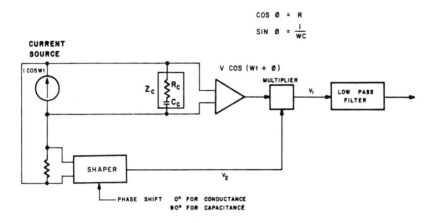

Fig. 7.4 Schematic diagram of a measuring circuit for capacitance, conductance, or impedance.

electrodes and connected to the instrument by means of a bottle basket assembly (Fig. 7.7).

Each BPU contains two separate air driven incubators, both capable of maintaining temperatures in the range of 10.0°C below ambient to 55.0°C. The cycle time, i.e. the time interval between consecutive readings for each sample, is 6 minutes or 0.1 hour.

The complete system includes the Bactometer Processing Units, a dedicated LSI-11/23 microcomputer equipped with 128K of memory, two eight inch double sided, double density floppy disk drives, a 19" (Intecolor) video display terminal, a fast printer (Anadex), and a digital plotter (Houston Instruments). The video terminal displays the system's menus and color-coded test results on the current status of each sample. A maximum of six impedance curves can be displayed simultaneously during or following completion of a test run. The printer is capable of rapidly providing permanent copies of stored information including product codes, sample assignments and detection time data. The digital plotter will plot up to six impedance curves simultaneously and, when operated in conjunction with the Bactometer's statistical program, is used to generate product calibration and statistical data.

The system is protected against transitory power failures of short duration such as may be expected to occur in production facilities. When power is restored, the system is automatically reactivated and data collection resumes. Extended power failures, however, can result in temperature fluctuations in the incubators which can adversely affect the growth rate of the microorganisms.

The stainless-steel electrodes employed, in conjunction with the 1.54 kHz driving source, enable the instrument to monitor changes in conductance, capacitance or total impedance. The considerations involved in choosing the best signal for a particular application are discussed in Chapter 4.6.

The system software includes on line algorithms which will automatically determine the detection time for each sample, and display continually updated color-coded results. Results appearing in red indicate that the corresponding detection time is earlier than a predetermined "cutoff" time, and that the sample contains a concentration of microorganisms in excess of the specified limit. Detections occurring after a "caution" time are displayed in green, indicating that these samples are "within spec". Results appearing in yellow indicate that detection occurred between the cutoff and caution times, suggesting that the samples are "marginal" from a microbiological point of view. See appendix A for the detailed statistical analysis.

The system was designed for simplicity of operation and rapid mastery by laboratory personnel with no previous experience in computer-controlled systems. By depressing one switch, all necessary systems are simultaneously activated. A single keystroke produces a list of all functions on the video screen. The program then guides the user into the correct procedure for initiating or terminating any of the computer's detection or display functions.

Fig. 7.5 The BACTOMETER® Microbial Monitoring System.

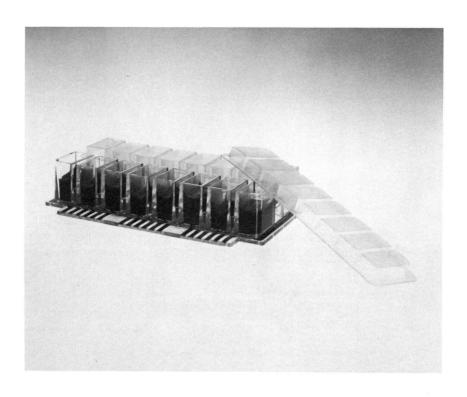

Fig. 7.6 A module containing 16 sample wells for use in the BACTOMETER® Microbial Monitoring System.

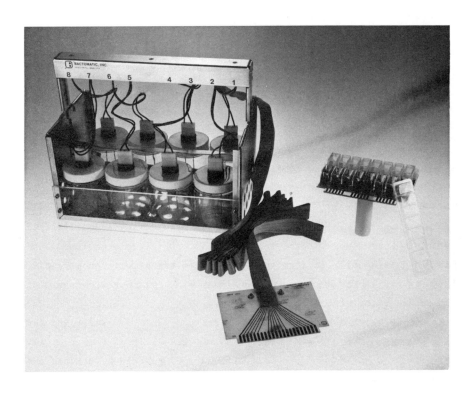

Fig. 7.7 Large vial bottle basket assembly for use in the BACTOMETER® Microbial Monitoring System.

Manufacturers of foods, drugs, cosmetics, etc. are constantly faced with the need to monitor a wide range of products and raw materials, each probably having different microbial specifications. The Bactometer® M123 system enables the operator to store such essential product information as sample type and selected test parameters. Once these parameters have been entered, the information is stored and will be automatically recalled when samples with the corresponding product code are labeled. Thus, when a product code is entered, the system automatically activates a specific sample detection routine to ensure reliable on line detection.

7.4 The Malthus Microbiological Growth Analysis*

*Section 7.4 is kindly provided by Malthus Instruments, Division of Matthey Printed Products Ltd., Stoke-on-Trent, U.K.

Malthus Microbiological Growth Analysers are manufactured by the Malthus Instrument Division of Matthey Printed Products Ltd., (Stoke-on-Trent, U.K.). Malthus systems provide a standardized automated method for the study of growth of micro-organisms. The hardware is modular in form to allow system expansion. An extensive library of software is available simplifying a wide range of processes in the microbiology laboratory.

Specific Malthus software programs are available for screening for micro-organisms in medicine (Baynes et al. 1983, Porter et al. 1983), food quality control (Gibson and Ogden 1980, McMurdo and Whyard, 1984), and the water industry. Other specialized needs are met by adaptations of these programs. The instrument design is based upon the work of Richards et al. (1978). Detection of yeasts, bacteria and other microorganisms is achieved by measuring the change in the conductance component of the electrical impedance.

A range of models is available, from an eight channel, low cost system, up to 256 channels under microcomputer control. A large analyser is shown in Figure 7.8.

Samples with low bacterial loadings, e.g. sterile food products, are best monitored with large samples which are accomodated in 100 ml cells. These are provided with a sample port for aseptic inoculation or sub-culture. Samples with high loadings, e.g. foodstuffs, may be investigated in smaller cells containing 10, 2 or 1 ml. Both types are re-usable glass cells fitted with robust platinum electrodes (Evans 1983).

The growth cells are housed in precision water heated/cooled incubators which may operate in the range +5 to 45°C. In high count applications, up to 128 individual cells are mounted on racks of eight cells as shown in Figure 7.9. In low count applications an incubator will accept up to 28 cells at any time, as shown in Figure 7.10. Data gathering is initiated automatically on insertion. Combinations of up to 256 cells of 2, 10 and 100 ml size may be monitored simultaneously. Data acquisition is undertaken from the control console which is protected against power failure by an uninterruptable power supply. The console control electronics multiplexes solid state switches to connect each cell to the measuring system. It also interfaces the resultant analog signals to the micro-computer which is fitted with a high resolution monitor. A graphics printer is provided for high speed output of alphanumeric or graphical data.

The system is controlled by disk resident software. This functions in two modes: "background", during which the data stream is received and processed and "foreground", during which the user interacts with the system.

In "background", data is processed into conductance change and stored on disk; user selected detection routines check this data base continuously for significant changes indicative of growth.

In "foreground", incubator status can be checked at a glance for cells showing positive detection, unallocated channels, etc. Data may be displayed in text or graphics form and permanent copy obtained using the graphics printer. It may also be selectively stored on disk for future reference. The software is menu driven and comprehensive HELP options aid rapid acceptance by inexperienced personnel. Extensive use is made of single keystroke commands for frequently used features. The modular software includes many features to ease laboratory management. Thus, search facilities are provided for recovery of current or archival data and data may be communicated to other computers.

Fig. 7.8 Malthus Microbiological Growth Analyser. Four incubator model for operation in the range +5 to 45°C.

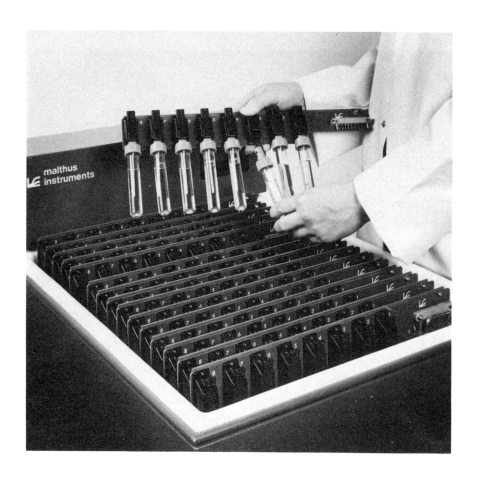

Fig. 7.9 Malthus High Count Incubator, general view showing 10 ml cells mounted in a cell rack together with cell racks loaded in the incubator. The Malthus High Count Incubator accepts 16 racks of 8 x 10 ml cells.

Fig. 7.10 Malthus Low Count Incubator; general view showing 28 100 ml growth cells together with electrical connections and sample ports.

APPENDIX A
Statistical Analysis of Scattergrams

STATISTICAL ANALYSIS

The following analysis is based upon the model developed in Chapter 3 and the first part of equation (3.6). The assumption is that detection time is linearly related to the logarithm of initial bacterial concentration:

$$\log C_{BO} = A' - B'(t_D - t_L)$$

We shall further assume that for a specific bacteria-medium-temperature combination the lag time t_L is constant. We can therefore combine constants in the above equation and disregard t_L.

$$\log C_{BO} = a - bt_D \qquad (A.1)$$

The experimental procedure deals with two variables pertinent to the sample: the initial bacterial concentration C_{BO} as determined by the plate count method, and the detection time t_D of the impedance method.

The first question to be addressed is the ability of the impedance method to correlate with the standard plate counts. Assuming that the chosen medium-temperature combination is appropriate for a specific application, the first step is to generate a scattergram of $\log C_{BO}$ versus t_D (see Fig. A.1). This is primarily a visual

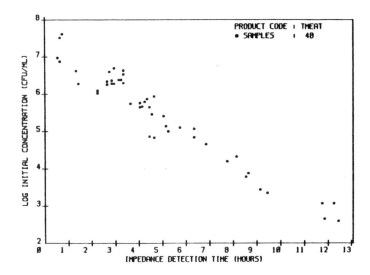

Fig. A.1 Scattergram of experimental data.

tool for identifying outlying points. These points can be reinvestigated to determine whether they resulted from errors associated with either method and whether they should be included in further analysis. The quantity which measures the correspondence between the two variables (the two methods in our case) is the correlation coefficient R. In Figure A.2 two scattergrams of the variables Y versus X are illustrated. The points obtained from methods that agree perfectly fall on a straight line and are characterized by a correlation coefficient magnitude of 1. The negative sign of R indicates a negative slope of the straight segment. The points from methods agreeing less satisfactorily are scattered, yielding a low correlation coefficient.

The correlation coefficient of the variables Y and X over N measurements points is given by:

$$R = \frac{\sum_{i=1}^{N} X_i Y_i - N\bar{X}\bar{Y}}{[(\sum_{i=1}^{N} X_i^2 - N\bar{X}^2)(\sum_{i=1}^{N} Y_i^2 - N\bar{Y}^2)]^{\frac{1}{2}}} \qquad (A.2)$$

Where X and Y are the mean values of the two variables X and Y.

$$\bar{X} = \frac{1}{N} \sum_{i=1}^{N} X_i$$

$$\bar{Y} = \frac{1}{N} \sum_{i=1}^{N} Y_i \qquad (A.3)$$

It should be emphasized that, although the correlation coefficient is also a measure of the dissimilarity between the methods, it does not indicate which of the methods is more responsible for less than perfect correlation.

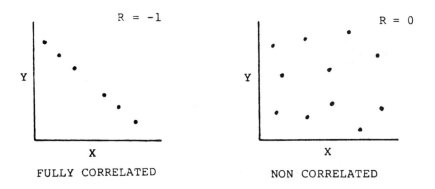

Fig. A.2   Illustration of correlation coefficients (R).

For the purposes of calibration, it is useful to find the slope and intercept of a straight line which best fit the scattergram. The linear regression technique is illustrated in Fig. A.3 where Yi and Xi are the coordinates of a single measured point. We wish to find the equation of a straight line which minimizes the sum of distances Di between the measured points and the corresponding points on the line, and which can eventually be compared with Equation (A.1). The line of best fit has the general form:

$$Y = A + BX \qquad (A.4)$$

where: $A = a$ and $B = -b$

The error, ERR, between the experimental points and the regression line is defined by:

$$ERR = \sum_{i=1}^{N} Di^2 = \sum_{i=1}^{N} (Y_i - \tilde{Y}_i)^2 = \sum_{i=1}^{N} (Y_i - A - BX_i)^2 \qquad (A.5)$$

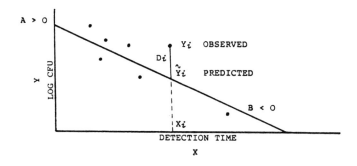

Fig. A.3 Linear fit to experimental data.

The minimum conditions for ERR are satisfied at the points where its partial derivatives with respect to A and B are zero:

$$\frac{\partial ERR}{\partial A} = 0$$

$$\frac{\partial ERR}{\partial B} = 0$$

(A.6)

The coefficients A and B can be obtained by resolving equations (A.6) and (A.5):

$$B = \frac{\sum_{i=1}^{N} X_i Y_i - N\bar{X}\bar{Y}}{\sum_{i=1}^{N} X_i^2 - N\bar{X}^2}$$

(A.7)

$$A = \bar{Y} - \bar{B}X$$

where: $\bar{X} = \frac{1}{N} \sum_{i=1}^{N} X_i$ and $\bar{Y} = \frac{1}{N} \sum_{i=1}^{N} Y_i$

Fig. A.4 is an example of a scattergram containing 25 actual data points. The correlation coefficient is -0.96, indicating high agreement between the impedance and the standard methods, and a regression line with negative slope. The equation of the regression line is:

$$\log C_{BO} = 8.61 - 0.28 \, t_D$$

The interpretation of the slope and the intercept of the regression line is discussed in Chapter 3.

Another question to be addressed is the width of the grey zone of the scattergram. For any product, there is usually a limit to the permissible concentration of bacteria. Since this concentration can now be expressed as a detection time, we can expect shorter detection times to indicate bacterial concentrations exceeding the permitted level, and longer times to indicate an acceptable product. The detection time which corresponds to the permissible limit is called permissible detection time (PDT). Since some of the scatter results from random errors of both methods, a grey zone should be created to ensure that points falling below the above limit do not actually belong to the area above it and vice versa. The width of this grey zone will depend on the application. It is possible, however, to suggest its width from further statistical examination of the data.

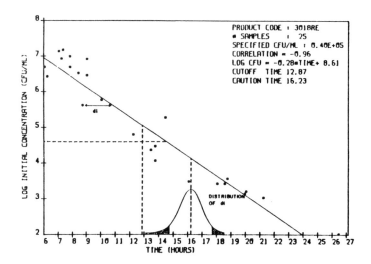

Fig. A.4 Linear regression analysis of experimental data.

Consider the example illustrated in Fig. A.4. The permissible level of microorganisms for this product (specified CFU/ml) is $4 \cdot 10^4$/ml. This concentration translates to a permissible detection time of 14.5 hours. Suppose that the detection time for a test sample is 14.6 hours. How confident can we be that this sample is acceptable? The answer might be that, since the detection time of the sample indicates contamination so close to the PDT, further checking of this particular sample is needed. The grey zone establishes a lower boundary (defined as cutoff time) and an upper boundary (defined as caution time) such that if a data point falls before the cutoff time there is only a slight chance that it actually belongs to the area beyond the caution time, and vice versa. The next assumption is a pessimistic one, in the sense that the plate count method is considered as the ideal and accurate method and that all the scatter results from the impedance method. In that case, the horizontal distances di (see Fig. A.4) are the major factors affecting the width of the grey zone. The standard deviation of these distances is given by:

$$SD = [\frac{1}{N} \sum_{i=1}^{N} di^2]^{\frac{1}{2}} \qquad (A.8)$$

In terms of line coefficients:

$$SD = [\frac{1}{N} \sum_{i=1}^{N} (Xi - \frac{1}{B}Yi + \frac{A}{B})]^{\frac{1}{2}} \qquad (A.9)$$

Now if we assume a normal distribution of the distances di and define cutoff time as the permissible detection time minus SD and caution time as the permissible detection time plus SD, then the distance between the caution time and the cutoff time is two standard deviations of the distribution of di. By the statistical hypothesis test (Bendat and Piersol, 1971) the width of the grey zone is such that a point falling exactly at the cutoff time has less than 5% probability of actually belonging to the area above the caution time, and vice versa. The probability that a point falling well below the cutoff time is actually misplaced is, of course, much lower than 5%. The probability is determined by the area of the normal distribution.

Let us evaluate how many samples will actually fall in the grey zone if it has the suggested width. This is an important matter since it determines what proportion of samples will have to be rechecked.

Fig. A.5 illustrates the scattergram generated for a particular application, together with the grey zone calculated as suggested. The distribution of points in the scattergram does not reflect the natural distribution of samples because sufficient "abused" samples were included to provide calibration over a large range of bacterial concentrations. The natural distribution is assumed to be normal with a mean detection time somewhere above caution time. Suppose that 5% of samples do not meet the specified level of 10^6/ml. The cutoff and caution times corresponding to this level of contamination are 2.57 and 4.17 hours respectively. The normal distribution of the samples, determined by the mean value and the 5% deviation, is superimposed on the scattergram. Since 5% of a normal distribution falls within the tail formed by 1.64 of a standard deviation, this number is indicated on the horizontal axis in the middle of the grey zone. Now, the relative area bounded by the tail of the distribution and the grey zone is the relative number of samples which need to be rechecked. Using the regression line equations we find that the cutoff time corresponds to 1.89 standard deviations of the distribution and the caution time corresponds to 1.39 standard deviations. From the appropriate statisical table of areas under a normal distribution we find tail areas of 2.94% for the caution time and 8.23% for the cutoff time. The difference 5.29% is the area bounded within the grey zone indicating the percentage of the samples that will have to be rechecked.

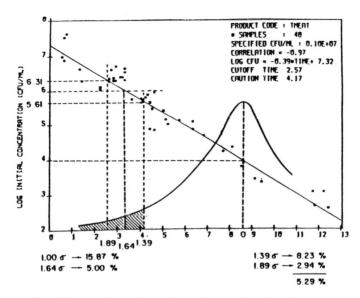

Fig. A.5 Illustration of number of samples in the grey zone.

In Section 3.4 the valid range of the electrobacteriological model was discussed. It was mentioned that although, within limits, the relation between IDT and log CFU/ml can be regarded as linear, we may expect nonlinear behavior for applications characterized by increased lag periods. For such applications, use of nonlinear regression techniques may result in a better data fit and higher correlation coefficients. Consider the scattergram illustrated in Fig. A.6 where it is desired to fit the data to the quadratic curve:

$$Y = A + BX + CX^2 \qquad (A.10)$$

where A, B and C are constants. Following the same procedure applied to the linear regression analysis, the sum of the distances di^2 between the observed values and the curve should be minimized. The constants can be calculated from the following determinants:

$$A = \frac{1}{\Delta} \begin{vmatrix} \Sigma y_i & \Sigma x_i & \Sigma x_i^2 \\ \Sigma x_i y_i & \Sigma x_i^2 & \Sigma x_i^3 \\ \Sigma x_i^2 y_i & \Sigma x_i^3 & \Sigma x_i^4 \end{vmatrix} \qquad B = \frac{1}{\Delta} \begin{vmatrix} N & \Sigma y_i & \Sigma x_i^2 \\ \Sigma x_i & \Sigma x_i y_i & \Sigma x_i^2 \\ \Sigma x_i^2 & \Sigma x_i^2 y_i & \Sigma x_i^4 \end{vmatrix}$$

$$C = \begin{vmatrix} N & \Sigma x_i & \Sigma y_i \\ \Sigma x_i & \Sigma x_i^2 & \Sigma x_i y_i \\ \Sigma x_i^2 & \Sigma x_i^3 & \Sigma x_i^2 y_i \end{vmatrix} \qquad \Delta = \begin{vmatrix} N & \Sigma x_i & \Sigma x_i^2 \\ \Sigma x_i & \Sigma x_i^2 & \Sigma x_i^3 \\ \Sigma x_i^2 & \Sigma x_i^3 & \Sigma x_i^4 \end{vmatrix}$$

The correlation coefficient R calculated in equation (A.2) was based on the assumption of a linear relationship between the two variables. The procedure can be extended to the nonlinear case by evaluating the correlation coefficient between the dependent variable Y and a new independent variable. Instead of using X (detection time) as the independent variable we may utilize the predicted value of Y as determined by X. Using the same formula (equation A.2) but replacing X by the predicted values, the R value is defined as the "multiple correlation coefficient". Note that the linear correlation coefficient is just a specific case of the more generalized multiple correlation.

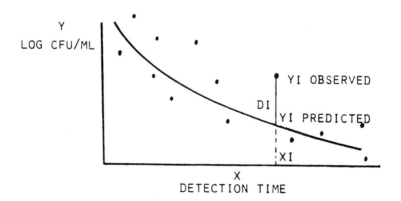

Fig. A.6 Quadratic curve fit to experimental data.

An example of a nonlinear regression analysis is illustrated in Fig. A.7. The quadratic fit in this case is superior to the linear approximation. This is indicated by its correlation coefficient of 0.90 compared with 0.87 for the linear fit.

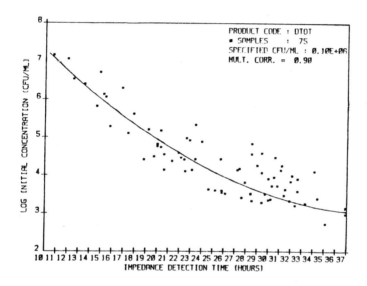

Fig. A.7 Quadratic regression analysis of experimental data.

APPENDIX B
Alphabetical List of Terms

VOCABULARY

ACCELERATION

The region in which rapid impedance changes take place. This is reflected on the impedance curves by an increasing slope relative to the slope of the baseline. The time at which acceleration is first observed is the impedance detection time.

ALGORITHM

An arithmetic formula for dealing with quantitative changes in impedance, permitting the instrument's software to automatically and reproducibly determine the impedance detection times of samples.

BASELINE

The region of the impedance curve between the end of the stabilization period and the onset of its acceleration phase.

BRIDGE

An electronic circuit to measure electrical properties (impedance, conductance, capacitance) in which the quantity being measured is compared with or balanced against a reference quantity.

CALIBRATION CURVE

The mathematical relation between two measured quantities. One form of the impedance calibration curve relates detection times to log initial concentration of microorganisms (CFU/ml). Other calibration curves may relate detection times to quantities such as shelf life. The calibration curve can be obtained by linear or quadratic regression.

CAPACITANCE

A quantity (measured in Farads or micro Farads) related to the capability of a device to store electrical energy. A capacitor typically consists of two plates separated by a dielectric that permits storage of energy in an electrical field between the plates. In an electrochemical well, capacitance results from the accumulation of charged ions at the interface between each electrode and the solution.

CAUTION TIME

A specific time chosen to ensure that growth detected after this point contains concentrations of microorganisms that are within the permitted limit. Its value depends on the permitted level (CFU/ml), the calibration curve, and the minimal confidence required in the classification.

CONDUCTANCE

The reciprocal of the resistance to a flow of electric current through a solution. Its value is determined by the concentration of mobile ions in the fluid. Solutions of minimal resistance have a high conductance, and vice versa. Conductance is measured in reciprocal ohms (mhos) or in Siemens units.

CORRELATION COEFFICIENT

A measure of the probability that a linear relationship exists between two observed quantities (e.g., log CFU/ml vs. detection time, or log CFU/ml vs. shelf life, etc.). The value of the coefficient ranges from 0 when there is no correlation, to ± 1, when there is complete correlation.

CUTOFF TIME

A specific time chosen to ensure that samples detected earlier contain microorganisms in excess of a specified limit. Its value depends on the permitted level (CFU/ml), the calibration curve, and the minimal statistical confidence required in the classification.

DETECTION

Observation of an acceleration in the impedance signal resulting from microbial growth and metabolism.

DETECTION TIME

See Impedance Detection Time.

DRIFT

The relative change of the impedance curve during its "baseline" phase, prior to the onset of acceleration. Ideally the drift is zero, i.e. the impedance does not change. In practice it usually drifts upwards or downwards. Drift is measured as percent change per hour.

ELECTRODE

A pair of metal wires or plates immersed in conductive growth medium and serving as the impedance sensor.

EQUILIBRATION TIME

See Stabilization Time.

GENERATION TIME

The time required for a population of microorganisms to double. It depends on the organism, the growth medium, and incubation temperature.

GREY ZONE

An interval on the calibration curve in which detected samples cannot be accurately classified with respect to the permitted concentration of microorganisms. Its width depends on the scatter of the data points around the calibration curve, and the level of statistical confidence (usually not less than 95%) required in the classification. The lower boundary of the grey zone is called the "cutoff time" and the upper boundary the "caution time".

IDT

See Impedance Detection Time.

IMPEDANCE

The total opposition to the flow of an alternating current. Impedance (Z) is the vector sum of resistive and reactive components, referred to in this book as the "G" and "C" signals. The G (conductive) signal results from changes occurring in solution, while the C (capacitive) signal results from changes in the electrically charged double layer at the electrode surface.

IMPEDANCE DETECTION TIME (IDT)

The time interval between the start of impedance monitoring and the beginning of the acceleration phase of the signal (conductance, capacitance, or impedance). The impedance detection time of a sample depends upon a number of parameters, including the initial concentration of microorganisms, the duration of their lag phase, and their generation time. The lag and generation times depend, in turn, on factors such as growth medium composition and incubator temperature.

LINEAR REGRESSION ANALYSIS

A mathematical procedure used to fit a straight line to data points. In our case each point represents a microbial concentration (CFU/ml) and its corresponding detection time.

LOGARITHM

Throughout the text the term log refers to the logarithm on the basis of 10. The term ln refers to the natural logarithm (basis of e).

MAXIMUM

The relative change of the impedance curve between detection time and time when maximum impedance level is attained.

METABOLISM COEFFICIENT

A quantity reflecting the activity of a single organism for a specific medium-temperature combination. It is defined as the number of ions generated by the organism per minute.

MODULE

Rectangular plastic sample holders designed for use with the Bactometer. Each module contains 16 individual sample wells arranged in two separate rows. Each sample well contains a pair of small stainless steel electrodes attached to a lead frame molded into the module's plastic base. Sample suspensions are pipetted into the appropriate test wells which are then covered with sterile plastic caps or sterile mylar tape. Filled modules are loaded into the Bactometer incubator by inserting the special metal connecting edge into the appropriate slot to complete the electrical connection.

PERMISSIBLE DETECTION TIME

Detection time of the impedance assay which corresponds to the specified permissible level.

PERMISSIBLE LEVEL/SPECIFIED PERMISSIBLE LEVEL

Refers to the maximum concentration of microorganisms allowed by specification, i.e. levels of bacteria above the permissible level are unacceptable.

QUADRATIC REGRESSION ANALYSIS

A mathematical procedure used to fit a quadratic curve (parabola) to data points. In our case each point represents a microbial concentration (CFU/ml) and its corresponding detection time.

SCATTERGRAM

A comparative plot of data from two measurement procedures which are assumed to be independent. In this book the variables are log CFU/ml and detection time.

SLOPE

The relative change (percent per hour) of an impedance curve established at the end of its acceleration. On a calibration curve, slope is the coefficient B of the straight line:

LOG CFU/ml = A + B · TIME where A is the line intercept.

SPECIFIED PERMISSIBLE LEVEL

See permissible level.

STABILIZATION TIME

The period required for the impedance signal to stabilize to its baseline, after the sample is placed into the incubator. Stabilization time depends on the temperature difference between sample and incubator, the volume of the well, and the medium-electrode electrochemical properties.

THRESHOLD

The minimum concentration of organisms (CFU/ml) at detection. The threshold for most microorganisms is approximately 10^7 viable organisms/ml.

WELL

A vial in which two electrodes are embedded. The growth medium added to the well completes the electrochemical circuit required for impedimetric measurements.

References

Allison, J.B., Anderson, J.A. and Cole, W.H. (1938). The method of electrical conductivity in studies of bacterial metabolism. J. Bacteriol. 36, 571-586.

Baynes, N.C., Comrie, J. and Prain, J.H. (1983). Detection of bacterial growth by the Malthus conductance meter. Med. Lab. Sci. 40, 149-158.

Beezer, A.E., Bitlelheim, K.A., Al-Salihi, S. and Shaw, E. (1978). The enumeration of bacteria in culture media and clinical specimens of wine by microcalimetry. Science Tools 25, 6-8.

Bishop, J.R., White, C.H. and Firstenberg-Eden, R. (1984). A rapid impedimetric method for determining the potential shelf-life of pasteurized whole milk. J. Food Prot. 47, 471-475.

Blankenagel, G. (1976). An examination of methods to assess post-pasteurization contamination. J. Milk Food Technol. 39, 301-304.

Bossuyt, R.G. and Waes, G.M. (1983). Impedance measurements to detect post-pasteurization contamination of pasteurized milk. J. Food Prot. 46, 622-624.

Brodsky, M.H. (1982); in Rapid Methods and Automation in Microbiology. (Edited by R.C. Tilton). p. 68-71, American Society of Microbiology, Washington, D.C.

Brown, D.F.J., Warner, M., Taylor, C.E.D. and Warren, R.E. (1984). Automated detection of micro-organisms in blood cultures by means of the Malthus Microbiological Growth Analyser. J. Clin. Pathol. 37, 65-69.

Cady, P. (1975); in New Approaches to the Identification of Microorganisms. (Edited by C.G. Heden and Illeni). p. 73-99, John Wiley & Sons, New York.

Cady, P. (1976). Detection and characterization of microorganisms by continuous impedance measurements. (Unpublished Data).

Cady, P. (1978); in Mechanizing Microbiology. (Edited by A.N. Sharpe, D.S. Clark and A. Balows). p. 199-239, Charles C. Thomas Pub. Co., Springfield.

Cady, P., Dufour, S.W., Lawless, P., Nunke, B. and Kraeger, J. (1978A). Impedance screening for bacteriuria. J. Clinic. Microbiol. 7, 273-278.

Cady, P., Hardy, D., Martins, S., Dufour, S.W. and Kraeger, S.J. (1978B). Automated impedance measurements for rapid screening of milk microbial content. J. Food Prot. 41, 277-283.

Colvin, H.J. and Sherris, J.C. (1977). Electrical impedance measurements in the reading and monitoring of broth dilution susceptibility tests. Antimicrobial Agents and Chemotherapy 12, 61-66.

Easter, M.C., Gibson, D.M. and Ward, F.B. (1982). A conductance method for the assay and study of trimethylamine oxide reduction. J. Appl. Bacteriol. 52, 357-365.

Eden, G. and Firstenberg-Eden, R. (1984). Enumeration of microorganisms by their AC conductance patterns. IEEE Trans. Biomed. Eng. 31, 193-198.

Edwards, P.R. and Ewing, W.H. (1972) in Identification of Enterobacteriaceae. Burgess Publishing Co., Minneapolis.

Evans, W.D.J. (1983). The measurements of bacterial growth. Platinum Metals Review. 27, 65.

Firstenberg-Eden, R. (1983). Rapid estimation of the number of microorganisms in raw meat by impedance measurements. Food Technol. 37, 64-70.

Firstenberg-Eden, R. (1984). A collaborative study of the impedance method for examining raw milk samples. J. Food Prot. Accepted for publication.

Firstenberg-Eden, R. and Klein, C.S. (1983). Evaluation of a rapid impedimetric procedure for the quantitative estimation of coliforms. J. Food Sci. 48(4), 1307-1311.

Firstenberg-Eden, R. and Tricarico, M.K. (1983). Impedimetric determination of total, mesophilic and psychrotrophic counts in raw milk. J. Food Sci. 48, 1750-1754.

Firstenberg-Eden, R., VanSise, M.L. and Klein, C.S. (1983). An impedimetric method for the presumptive identification of Salmonella. IFT 1983 Abstract.

Firstenberg-Eden, R., VanSise, M.L., Zindulis, J. and Kahn, P. (1984). Impedimetric coliform estimation in dairy products. Submitted for publication.

Firstenberg-Eden, R. and Zindulis, J. (1984). Electrochemical changes in media due to microbial growth. J. Microbiol. Methods. 2, 103-115.

Fisher, R.A., Thornton, H.G. and MacKenzie, W.A. (1922). The accuracy of the plating method of estimating the density of bacterial population. Ann. Appl. Biol. 9, 325-359.

Gibson, D.M. and Ogden, I.D. (1980). Assessing bacterial quality of fish by conductance measurements. J. Appl. Bacteriol. 49, 12.

Gilchrist, J.E., Donelly, C.B., Peeler, J.T. and Delany, J.M. (1973). Spiral plate method for bacterial determination. Appl. Microbiol. 25, 244-252.

Gnan, S. and Luedecke, L.O. (1982). Impedance measurements in raw milk as an alternative to the standard plate count. J. Food Prot. 45, 4-7.

Hadley, D., Kraeger, S.J., Dufour, S.W. and Cady, P. (1977). Rapid detection of microbial contamination in frozen vegetables by automated impedance measurements. Appl. Environ. Microbiol. 34, 14-17.

Hadley, D., and Senyk, G. (1975). Early detection of microbial metabolism and growth by measurement of electrical impedance. Microbiology 1975. p. 12-21. American Society for Microbiology, Washington D.C.

Hankin, L. and Anderson, E.O. (1969). Correlation between flavor score, flavor criticism, standard plate count, and oxidase count on pasteurized milks. J. Milk Food Technol. 32, 49-51.

Hankin, L., and Stephens, G.R. (1972). What tests usefully predict keeping quality of perishable foods? J. Milk Food Technol. 35, 574-576.

Hankin, L., Dillman, W.F. and Stephens, G.R. (1977). Keeping quality of pasteurized milk for retail sale related to code date, storage temperature and microbial counts. J. Food Prot. 40, 848-853.

Hammond, S.M. and Carr, J.C. (1976); in *Inhibition and Inactivation of Vegetative Microbes*. (Edited by F.P. Skinner and N.B. Auzo). p. 89-110. Society of Appl. Bacteriology Symposia Series #51.

Hause, L.L., Komorowski, R.A. and Gayon, F. (1981). Electrode and electrolyte impedance in the detection of bacterial growth. IEEE Transaction. Biomed. Eng. 28, 403-410.

Hobbs, G. and Gibson, D.M. (1977). The use of conductance for measuring microbiological growth rates. J. Applied Biology. 43, 3.

Holland, R.L., Cooper, B.H., Helgeson, H.G.P. and McCracken, A.W. (1980). Automated detection of microbial growth in blood cultures by using stainless-steel electrodes. J. Clinical Microbiol. 12, 180-184.

International Commission on Microbiological Specification for Foods (ICMSF) (1978); in *Microorganisms in Foods*. (2nd Edition). p. 137. Univ. of Toronto Press, Toronto.

International Commission on Microbiological Specification for Foods (ICMSF). (1978B); in *Microorganisms in Foods* (Part 2); Sampling for Microbiological Analysis:Principles and Specific Applications, (2nd Edition). p. 6-63. Univ. of Toronto Press, Toronto.

Irvin, R.T., MacAlister, T.J., Chan, R. and Costerton, J.W. (1981). Citrate-Tris (hydroxymethyl) aminomethane-mediated release of outer membrane sections from the cell envelope of a deep-rough (heptose deficient lipopolysaccharide) strain of *Escherichia coli* 08. J. Bacteriol. 45, 1386-1396.

Jason, A.C. (1983). A deterministic model for monophasic growth of batch cultures of bacteria. Antonie van Leeuwenhoek. 49, 513-536.

Johnston, H.H. and Newsom, S.W.B. (Ed). (1976). Rapid methods and automation in Microbiology (2nd International Symposium). Learned Information (Europe) Ltd., Oxford and New York.

Kagan, R.L., Schuette, W.H., Zierdt, C.H. and MacLowry, J.D. (1977). Rapid automated diagnosis of bacteremia by impedance detection. J. Clin. Microbiol. 5, 51-57.

Kahn, P. and Firstenberg-Eden, R. (1984). A new cosmetic sterility test. Soap, Cosmetics and Chemical Specialties. 60, 46-48; 101.

Khan, W., Friedman, G. Rodriguez, W., Controni G. and Ross, S. (1976). Rapid detection of bacteria in blood and spinal fluids in children by electrical impedance method. Paper C-70, American Society for Microbiology 77th Annual Meeting.

Kallings, L.O. (1967). Sensitivity of various Salmonella strains to Felix-01-Phage. Acta. Pathol. Microbiol. Scand. 70, 446-454.

Lamb, V.A., Dalton, H.P. and Wilkins, J.R. (1976). Electrochemical method for the early detection of urine-tract infections. Ann. J. Clin. Pathol. 66, 91-95.

Martins, S.B., Hodapp, S., Dufour, S.W. and Kraeger, S.J. (1982). Evaluation of a rapid impedimetric method for determining the keeping quality of milk. J. Food Prot. 45, 1221-1226.

McMurdo, I.H. and Whyard, S. (1984). Suitability of rapid microbiological methods for the hygienic management of spray drier plant. J. Soc. Dairy-Tech. 1, 4-9.

McPhillis, J. and Snow, N. (1958). Studies on milk with a new type of conductivity cell. Australian J. Dairy Technol. 3, 192-196.

Mott, N.F. and Watts-Tobin, R.J. (1961). The interface between a metal and an electrolyte. Electrochem. Acta 4, 79-107.

Niemala, S. (1983). Statistical evaluation of results from quantitative microbial examinations. Nordic Committee On Food Analysis, Report #1 (2nd Edition).

O'Connor, F. (1979). An impedance method for the determination of bacteriological quality of raw milk. IR. J. Fd. Sci. Technol. 3, 93-100.

Oker-Blom, M. (1912). Die elecktrische Leitfahigkeit im dienste der bacteriologie. Centralbl. J. Bakteriol. Abs. 65, 382-389.

Parsons, L.B., Drake, E.T. and Slurges, W.S. (1929). A bacteriological conductivity culture cell and some of its applications. J. Amer. Chem. Soc. 51, 166-171.

Parsons, L.B. and Sturges, W.S. (1926A). The possibility of the conductivity method as applied to studies of bacterial metabolism. J. Bacteriol. 11, 177-188.

Parsons, L.B. and Sturges, S.W. (1926B). Conductivity as applied to studies of bacterial metabolism. J. Bacteriol. 11, 267-272.

Pettipher, G.L., Mansell, R., McKinnon, C.H. and Cousin, C.M. (1980). Rapid membrane filtration-epiflourescent microscopy technique for direct enumeration of bacteria in raw milk. Appl. Environ. Micro. 39, 423-429.

Porter, I.A., Reid, T.M.S., Wood, W.J., Gibson, D.M. and Hobbs, G. (1983); in Antibiotics:Assessment of Antimicrobial Activity and Resistance. (Edited by A.D. Russell and L.B. Quesnel). p. 49-60, Academic Press, London.

Randolph, H.E., Freeman, T.R. and Peterson, R.W. (1965). Keeping quality of market milk obtained at retail outlets and at processing plants. J. Milk Food Technol. 28, 92-96.

Rantama, A. (1983). Impedimetric detection of Staphylococci. Private communication.

Richards, J.C.S., Jason, A.C., Hobbs, G., Gibson, D.M. and Christie, R.H. (1978). Electronic measurement of bacterial growth. J. Phys. E.:Sci. Instrum. 11, 560-568.

Schindler, P.R.G., and Tueber, M. (1978). Ultrastructural study of Salmonella typhimurium treated with membrane-active agents: specific reaction of dansylchloride with cell envelope components. J. Bacteriol. 135, 198-206.

Schwan, H.P. (1963); in *Physical Techniques in Biological Research* (Volume VI); Electrophysiological Methods, Part B. Academic Press Inc., New York.

Schwan, H.P. (1966). Alternating current electrode polarization. Biophysik 3, 181-201.

Sharpe, A.N. (1979). Germ of a new food microbiology. New Scientist, 13, 860-862.

Sharpe, A.N. (1979B). An alternative approach to food microbiology for the future. Food Technol. 33(3), 171-74.

Sharpe, A.N. (1980); in *Food Microbiology - A Framework for the Future*. Charles C. Thomas, Springfield.

Sogaard, H. and Lund, R. (1981); in *Psychrotrophic Microorganisms in Spoilage and Pathogenicity*. p. 91. Academic Press Inc., New York.

Spector, W.B. (ed.) (1956). Handbook of Biological Data. Table 75. Saunders, Philadephia, PA.

Spector, S., Throm, R., Strauss, R. and Friedman, H. (1977). Rapid detection of bacterial growth in blood samples by a continuous-monitoring electrical impedance apparatus. J. Clinical Microbiol. 6, 489-493.

Stewart, G.N. (1899). The change produced by the growth of bacteria in the molecular concentration and electrical conductivity of culture media. J. Exp. Med. 4:235-245.

Stewart, B.J., Eyles, M.J. and Murrell, W.G. (1980). Rapid radiometric method for detection of *Salmonella* in foods. Appl. Environ. Microbiol. 40, 223-230.

Suomalainen, H. and Oura, E. (1971); in *The Yeasts* (Volume 2). (Edited by A.H. Rose and J.S. Harrison). p. 3-60. Academic Press Inc., New York.

Throm, R., Specter, S., Strauss R. and Friedman, H. (1977). Detection of bacteriuria by automated electrical impedance monitoring in a clinical microbiological laboratory. J. Clinical Microbiol. 6, 271-273.

Throm, R., Strauss, S., Specter, S. and Friedman, H. (1976). Automated blood culture testing using radioisotope and electrical impedance monitoring equipment. Second International Symposium of Rapid Methods and Automation in Microbiology.

Tilton, R.C. (ed.) (1982). Rapid methods and automation in microbiology. Proceedings of the Third International Symposium on Rapid Methods and Automation in Microbiology. American Society for Microbiology, Washington, DC.

Ur, A. and Brown, D.F.J. (1975); in, New Approaches to the Identification of Microorganisms. (Edited by C.G. Heden and T. Illeni). p. 61-71. John Wiley and Sons, New York.

Velz, C.J. (1951). Graphical approach to statistics. IV. Evaluation of bacterial diversity. Water & Sewage Work 98, 66-74.

Warburg, E. (1899). Über das verhalten sogenannter unpolarisirbarer electroden gegen Wechselstrom. Ann. Phy. Chem. 67, 493-499.

Warburg, E. (1901). Über die Polarization capacitat des platins. Ann. Phys. 6, 125.

Watrous, G.H., Jr., Barnard, S.E. and Coleman, W.W. (1971). A survey of the actual and potential keeping quality of pasteurized milk from 50 Pennsylvania dairy plants. J. Milk Food Technol. 34, 145-149.

Welkos, S., Schreiher, M. and Baer, H. (1974). Identification of Salmonella with 0-1 Bacteriophage. Appl. Microbiol. 28, 618-622.

Wheeler T.G. and Goldschmidt, M.C. (1975). Determination of bacterial cell concentration by electrical measurements. J. Clin. Microbiol. 1, 25-29.

Wickerham, L.J. (1951). Taxonomy of Yeasts. Tech. Bull. U.S. Dept. Agriculture No. 1029.

Wilkins, J.R., Stoner, G.E. and Baykin, E.H. (1974). Microbial detection method based on sensing molecular hydrogen. Appl. Microbiol. 27, 949-952.

Wood, A.J. and Baird, E.A. (1943). Reduction of trimethylamine oxide bacteria. 1. The <u>Enterobacteriaceae</u>. J. Fisheries Res. Board Can. <u>6</u>, 194-201.

Wood, J.M. and Gibbs, P.A. (1982); in, <u>Developments in Food Microbiology - 1</u>. (Edited by R. Davis).

Zall, R.R., Chen, J.H., Murphy, S.C. (1982). The detection of psychrotrophic bacteria in milk using a dye test. Cult. Dairy Prod. J. <u>17</u>, 7-12.

Zafari, Y. and Martin, W.J. (1977). Comparison of the Bactometer® Microbial Monitoring System with conventional methods for detection of microorganisms in urine. J. Clinical Microbiol. <u>5</u>, 545-547.

Ziedt, C.H., Kagan, R., and MacLowry, J.D. (1977). Development of a lysis-filtration blood culture technique. J. Clin Microbiol. <u>5</u>, 46-50.

Zindulis, J. (1984A). A medium for the impedimetric detection of yeasts in foods. Food Microbiology. In press.

Zindulis, J. (1984B). The impedimetric detection of yeast in yogurt. ADSA Abstract.

Index

Abused samples 80, 81
Acceleration 17, 45, 151
Accuracy,
  impedance 76, 77
  plate count 74-76
Activity,
  microbial 26, 32, 38, 120, 121
Admittance 11
Antibiotics 107-112

Bioluminescence 4
Blood 4, 100, 103-106
Bridge 9, 125-128, 151

Calibration curve 74, 77-84, 153
<u>Candida</u> <u>utilis</u> 60, 61, 63, 64
Capacitance 9-11, 13-16, 21, 22, 57-64, 126-128, 154
Characterization of organisms 113-121
Coliforms 51, 52, 92, 93
Conductivity 9-13, 21, 22, 57-64, 126-128, 154
Confidence limits 5, 146, 147
Correlation coefficient 77, 78, 142, 143, 150, 154
Cosmetics 100, 106, 107
Curves 43, 46-52

Detection time 17-20, 27, 28, 46, 154, 156
Drift 44, 45, 155

<u>E. coli</u> 34, 42, 56, 57, 60, 61, 90, 104, 109, 110, 118
Electrobacteriological model 21, 23-39
Electrode 10, 11, 13-16, 19, 58, 155
Enumeration 2, 3, 73-98
Epifluorescence 3
Errors,
  sampling 29, 70-71, 76
  manipulation 77
Fish 88
Frequency 57-58
Generation time 19, 25, 26, 31, 52-57, 76, 83, 84, 120, 155
Growth kinetics 120, 121

Hydrophobic grid membrane filter 1, 2

Identification of organisms 2, 113-121
Inhibitors 2, 67-70
Injured organisms 2, 42, 43, 66, 67

169

Kloeckera apiculata 97

Linear regression 144,-145

Meat 87, 88
Media,
  effect on impedance 46-49, 58, 59
Metabolism coefficient 26, 32, 38
Microcalorimetry 3, 4
Minimal inhibitory concentration (MIC) 107-112
Milk 84-87, 92, 93, 100-103, 114, 121
Moseley test 102, 103
Most probable number 2

Neisseria Gonorrhoeae 113

Orange juice 97

pH 62-64
Phage sensitivity 117-119
Preincubation 72
Proteus 90
Pseudomonas 48, 49, 54, 55, 59, 90, 104

Quadratic regression 149-151, 157
Quality control 81-84
Quantitation 2, 3, 73-98

Radiometry 4
Rapidity 32-33, 39
Regression analysis 5, 77-89, 141-151

Saccharomyces cerevisiae 96
Salmonella,
  identification 114-119
Sampling error 29, 70, 71, 76
Selective media 90-98

Sensitivity to antibiotics 107-112
Shelf life 5, 99-103
Spiral plater 1
Sterility test 103-107
Susceptibility tests, 2, 107-112

Temperature 19, 20, 42, 55-57
Threshold,
  bacterial 17, 33, 34, 38, 157
  ionic 17

Urine 3, 8, 89, 90, 112

Vegetables,
  frozen 88

Yeasts 59, 60, 63, 93-98
Yogurt 97-98